VECTOR ANALYSIS
AND CARTESIAN TENSORS

VECTOR ANALYSIS
AND
CARTESIAN TENSORS

SECOND EDITION

D. E. BOURNE
University of Sheffield

and

P. C. KENDALL
University of Keele

VNR

International

To Brenda and Hazel

First published in 1967 by
Van Nostrand Reinhold (International) Co. Ltd
11 New Fetter Lane, London EC4P 4EE

Second edition 1977
Reprinted 1982, 1983, 1984, 1985, 1986, 1988, 1990

© 1967, 1977 D.E. Bourne and P.C. Kendall

Printed in Hong Kong

ISBN 0 442 30743 8

PREFACE

The most popular textbook approach to vector analysis begins with the definition of a vector as an equivalence class of directed line segments— or, more loosely, as an entity having both magnitude and direction. This approach is no doubt appealing because of its apparent conceptual simplicity, but it is fraught with logical difficulties which need careful handling if they are to be properly resolved. Consequently, students often have difficulty in understanding fully the early parts of vector algebra and many rapidly lose confidence. Another disadvantage is that subsequent developments usually make frequent appeal to geometrical intuition and much care is needed if analytical requirements are not to be obscured or overlooked. For example, it is seldom made clear that the definitions of the gradient of a scalar field and the divergence and curl of a vector field imply that these fields are continuously differentiable, and hence that mere existence of the appropriate first order partial derivatives is insufficient.

The account of vector analysis presented in this volume is based upon the definition of a vector in terms of rectangular cartesian components which satisfy appropriate rules of transformation under changes of axes. This approach has now been used successfully for ten years in courses given from the first year onwards to undergraduate mathematicians and scientists, and offers several advantages. The rules for addition and subtraction of vectors, for finding scalar and vector products and differentiation are readily grasped, and the ability to handle vectors so easily gives the student immediate confidence. The later entry into vector field theory takes place naturally, with gradient, divergence and curl being defined in their cartesian forms. This avoids the alternative more sophisticated definitions involving limits of integrals. Another advantage of the direct treatment of vectors by components is that introducing the student at a later stage to tensor analysis is easier. At that stage, tensors are seen as a widening of the vector concept and no mental readjustment is necessary.

The approach to vectors through rectangular cartesian components does not obscure the intuitive idea of a vector as an entity with magnitude and direction. The notion emerges as an almost immediate consequence of the definition, and is more soundly based inasmuch as both the magnitude and direction then have precise analytical interpretations. The familiar parallelogram law of addition also follows easily.

The essential background ideas associated with rotations of rectangular cartesian coordinate axes are introduced in Chapter 1 at a level suitable for undergraduates beginning their first year. The second and third chapters deal, respectively, with the basic concepts of vector algebra and differentiation of vectors; applications to the differential geometry of curves are also given in preparation for later work.

Vector field theory begins in Chapter 4 with the definitions of gradient, divergence and curl. We show also in this chapter how orthogonal curvilinear coordinate systems can be handled within the framework of rectangular cartesian theory.

An account of line, surface and volume integrals is given in the fifth chapter in preparation for the integral theorems of Gauss, Stokes and Green which are discussed in Chapter 6. The basic approach to vectors that we have adopted enables rigorous proofs to be given which are nevertheless within the grasp of the average student. Chapter 7 deals with some applications of vector analysis in potential theory and presents proofs of the principal theorems.

Chapters 8 and 9 on cartesian tensors have been added to this second edition in response to the suggestion that it would be useful to have between two covers most of the vector and tensor analysis that undergraduates require. The case for adding this material is strengthened by the fact that the approach to vectors in the early chapters makes the transition to tensors quite straightforward. Chapter 8 deals with the basic algebra and calculus of cartesian tensors, including an account of isotropic tensors of second, third and fourth order. Chapter 9 briefly discusses those properties of second order tensors which have risen to importance in continuum mechanics over the last twenty years. Some theorems on invariants and the representation of isotropic tensor functions are proved.

We acknowledge warmly the many useful comments from students and colleagues who have worked with the first edition. They have enabled us to make improvements to the original text. We particularly thank the following: Dr. G. T. Kneebone and Professor L. Mirsky for their early interest in the first edition, and Professor A. Jeffrey and Thomas Nelson and Sons Ltd. without whom this new edition would not have appeared.

D. E. BOURNE
P. C. KENDALL

CONTENTS

Chapter 4   Scalar and Vector Fields

Chapter 5   Line, Surface, and Volume Integrals

Chapter 6   Integral Theorems

Chapter 7   Applications in Potential Theory

CHAPTER 1

RECTANGULAR CARTESIAN COORDINATES
AND ROTATION OF AXES

1.1 Rectangular cartesian coordinates

From a fixed point O, which we shall call the *origin of coordinates*, draw three fixed lines Ox, Oy, Oz at right angles to each other, as in Fig. 1. These will be called the x-axis, y-axis, z-axis, respectively; and they will be

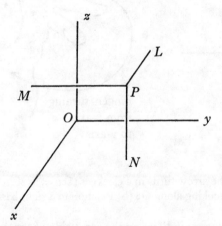

FIG. 1. Rectangular cartesian coordinates

referred to collectively as rectangular cartesian axes $Oxyz$. The planes Oyz, Ozx, and Oxy are called the coordinate planes, and they may be referred to as the yz-plane, zx-plane, and xy-plane respectively.

It is customary to choose the axes in such a way that Ox, Oy, and Oz form a right-handed set, in that order. This means that to an observer looking along Oz, the sense of the smaller arc from a point on Ox to a point on Oy is clockwise. Figure 2(a) illustrates this; Fig. 2(b) shows the relationship of Fig. 2(a) to a right hand. Notice that to an observer looking along Ox, the sense of the smaller arc from Oy to Oz is clockwise; and to an observer looking along Oy, the sense of the smaller arc from Oz to Ox is clockwise.

The three statements concerning observers looking along the respective

axes exhibit *cyclic symmetry in x, y, z*; i.e., if, in any one of the three statements, we replace x by y, y by z and z by x, then one of the other two statements is produced. The operation of replacing x by y, y by z and z by x is
called a *cyclic interchange of x, y, z*.

The position of a point P relative to a given set of rectangular cartesian
axes may be specified in the following way. Draw the perpendiculars, PL,
PM, PN from P to the yz-, zx-, xy-planes, respectively, as in Fig. 1. Let

$$x = \pm\text{length of } PL,$$

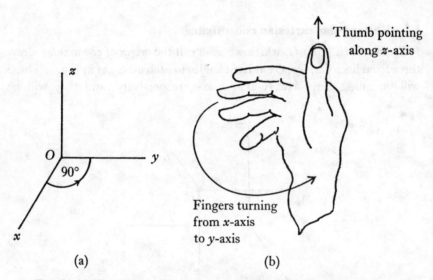

(a)   (b)

FIG. 2. (a) The arrow turns in a clockwise sense, as seen by an observer
looking along Oz (b) Relation to a right hand

taking the positive sign if P lies on the same side of the yz-plane as Ox, and
the negative sign otherwise. Similarly, define

$$y = \pm\text{length of } PM,$$
$$z = \pm\text{length of } PN,$$

the positive or negative sign for y being taken according as P lies on the
same or on the opposite side of the zx-plane as Oy, and the positive or
negative sign for z being taken according as P lies on the same or on the
opposite side of the xy-plane as Oz. The numbers x, y, z are called the
x-coordinate, y-coordinate, z-coordinate of P. We may refer to P as the
point (x, y, z).

It is an elementary observation that, when x, y, z are given, the position
of P relative to the given axes is determined uniquely. Conversely, a given
point P determines a unique triad of coordinates. In other words, there

is a *one–one correspondence* between points P and triads of real numbers (x, y, z).

Distance from origin. To find the distance of P from the origin O, construct the rectangular parallelepiped which has PL, PM, PN as three edges (Fig. 3). Using Pythagoras's theorem, we have

$$OP^2 = ON^2 + PN^2$$
$$= PL^2 + PM^2 + PN^2.$$

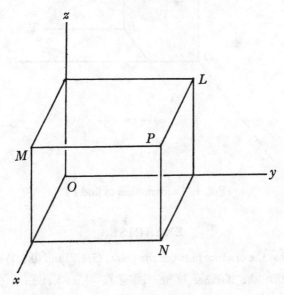

FIG. 3. Construction to find the distance OP

Since the perpendicular distances of P from the coordinate planes are $|x|$, $|y|$, $|z|$, it follows that

$$OP = \sqrt{(x^2 + y^2 + z^2)}. \tag{1.1}$$

Distance between points. The distance between the points $P(x, y, z)$ and $P'(x', y', z')$ may be found in the following way. Through P construct three new coordinate axes PX, PY, PZ parallel to the original axes Ox, Oy, Oz, as shown in Fig. 4. Let the coordinates of P' *relative to these new axes* be X, Y, Z. Then it is easily seen that

$$X = x' - x, \quad Y = y' - y, \quad Z = z' - z.$$

Applying result (1.1),

$$PP' = \sqrt{(X^2 + Y^2 + Z^2)},$$

and so, in terms of coordinates relative to the original axes,

$$PP' = \sqrt{\{(x'-x)^2+(y'-y)^2+(z'-z)^2\}}. \tag{1.2}$$

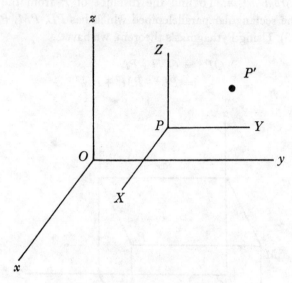

FIG. 4. Construction to find PP'

EXERCISES

1. Show that the distance between the points $(5,4,2)$ and $(0,3,1)$ is $3\sqrt{3}$.

2. Show that the distance of the point $(a-b, a+b, c)$ from the origin is $\sqrt{(2a^2+2b^2+c^2)}$.

3. Find the points in the xy-plane which are at unit distance from the origin and equidistant from the x-axis and y-axis.

4. Find the points which are at a distance of 5 units from the origin and whose distances from both the xy- and zx-planes are $2\sqrt{2}$ units.

5. Find the points which are at a distance of $\frac{1}{2}\sqrt{2}$ from every axis.

6. Find the distance between (i) the points $(1,-1,0)$ and $(1,2,4)$, (ii) the points $(3,-1,2)$ and $(-1,5,-1)$.

7. The coordinates of a point O' relative to rectangular cartesian axes Ox, Oy, Oz are $(1,1,-1)$. Through O', new axes $O'x'$, $O'y'$, $O'z'$ are taken such that they are respectively parallel to the original axes. Find the coordinates of O relative to the new axes. If a point P has coordinates $(-1,2,0)$ relative to the new axes, find its perpendicular distances from the xy-, xz-, yz-planes.

8. Find the length of the perimeter of the triangle whose vertices lie at the points $(1,0,0)$, $(0,1,0)$ and $(0,0,1)$.

1.2 Direction cosines and direction ratios

Direction cosines. Let OP be a line described in the sense from O (the origin) to a point P, and denote by α, β, γ the angles that OP makes with Ox, Oy, Oz (Fig. 5). We define the *direction cosines* of OP to be $\cos\alpha$, $\cos\beta$, $\cos\gamma$. For convenience we write

$$l = \cos\alpha, \quad m = \cos\beta, \quad n = \cos\gamma. \tag{1.3}$$

The direction cosines of the x-axis, for example, are 1, 0, 0.

Denote the foot of the perpendicular from P to the x-axis by N, let $OP = r$, and suppose that the coordinates of P are (x, y, z). From the triangle OPN

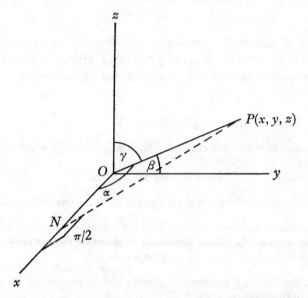

FIG. 5. The line OP makes angles α, β, γ with the axes

we have $ON = |x| = r|\cos\alpha|$. Also, if α is an acute angle, $\cos\alpha$ and x are both positive, whilst if α is an obtuse angle, $\cos\alpha$ and x are both negative. It follows that $x = r\cos\alpha$, and similarly we may show that $y = r\cos\beta$, $z = r\cos\gamma$. The direction cosines of OP are therefore

$$l = x/r, \quad m = y/r, \quad n = z/r. \tag{1.4}$$

Since $r^2 = x^2 + y^2 + z^2$, we have

$$l^2 + m^2 + n^2 = 1. \tag{1.5}$$

This shows that the direction cosines of a line are not independent: they must satisfy (1.5).

The direction cosines of a line $\mathcal{L}$ not passing through the origin aie defined to be the same as those of the parallel line drawn from the origin in the same sense as $\mathcal{L}$.

Direction ratios. Any three numbers a, b, c such that

$$a:b:c = l:m:n \tag{1.6}$$

are referred to as *direction ratios* of OP. If (1.6) holds, we have

$$l = a/d, \quad m = b/d, \quad n = c/d, \tag{1.7}$$

where, by substituting into equations (1.5),

$$d = \pm\sqrt{(a^2+b^2+c^2)}. \tag{1.8}$$

The choice of sign in (1.8) indicates that there are two possible sets of direction cosines corresponding to a given set of direction ratios. These sets of direction cosines refer to oppositely directed parallel lines.

EXERCISES

9. Show that the direction cosines of the line joining the origin to the point $(1, -4, 3)$ are

$$1/\sqrt{26}, \ -4/\sqrt{26}, \ 3/\sqrt{26}.$$

10. Find the direction cosines of the line joining the origin to the point $(6, 2, 5)$.

11. A line makes angles of $60°$ with both the x-axis and y-axis, and is inclined at an obtuse angle to the z-axis. Show that its direction cosines are $\frac{1}{2}$, $\frac{1}{2}$, $-\frac{1}{2}\sqrt{2}$, and write down the angle it makes with the z-axis.

12. Find the direction cosines of the line which is equidistant from all three axes and is in the positive octant $x \geqslant 0$, $y \geqslant 0$, $z \geqslant 0$.

13. Find direction ratios for the line which makes an angle of $45°$ with the x-axis and an angle of $45°$ with the y-axis, and lies in the positive octant.

1.3   Angles between lines through the origin

Consider two lines OA and OA', with direction cosines l, m, n and l', m', n'. To find the angle θ between them, denote by B, B' the points on OA, OA' (produced if necessary) such that $OB = OB' = 1$ (Fig. 6). Then the coordinates of B, B' are (l, m, n), (l', m', n'), from equation (1.4) with $r = 1$. Applying the cosine rule to the triangle OBB' gives

$$\cos\theta = \frac{OB^2 + OB'^2 - BB'^2}{2OB.OB'} = 1 - \tfrac{1}{2}BB'^2.$$

But from (1.2),

$$BB'^2 = (l'-l)^2+(m'-m)^2+(n'-n)^2$$
$$= (l'^2+m'^2+n'^2)+(l^2+m^2+n^2)-2(ll'+mm'+nn').$$

Using the results $l^2+m^2+n^2=1$, $l'^2+m'^2+n'^2=1$, we obtain

$$\cos\theta = ll'+mm'+nn'. \tag{1.9}$$

Note that, because $\cos(2\pi-\theta)=\cos\theta$, we still obtain equation (1.9) when the angle between OA and OA' is taken as $2\pi-\theta$.

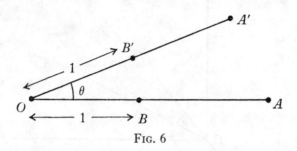

Fig. 6

Condition for perpendicular lines. Two lines through the origin are perpendicular if and only if

$$ll'+mm'+nn' = 0. \tag{1.10}$$

Proof. The two lines OA, OA' are at right angles if and only if $\theta=\frac{1}{2}\pi$ or $\theta=\frac{3}{2}\pi$, i.e. if and only if $\cos\theta=0$. The result now follows from (1.9).

EXERCISES

14. Show that the angle between the lines whose direction cosines are $\frac{1}{2}\sqrt{2}$, $\frac{1}{2}\sqrt{2}$, 0 and $\frac{1}{3}\sqrt{3}$, $\frac{1}{3}\sqrt{3}$, $\frac{1}{3}\sqrt{3}$ is $\cos^{-1}\frac{1}{3}\sqrt{6}$.

15. Show that the lines whose direction cosines are $\frac{1}{2}\sqrt{2}$, 0, $\frac{1}{2}\sqrt{2}$ and $\frac{1}{2}\sqrt{2}$, 0, $-\frac{1}{2}\sqrt{2}$ are perpendicular.

16. Find the angle between any two of the diagonals of a cube. [*Hint.* Choose axes suitably, with origin at the centre of the cube.]

1.4   The orthogonal projection of one line on another

Let two lines OP, OA meet at an angle θ. Then we define the orthogonal projection of OP on OA to be $OP\cos\theta$ (Fig. 7).

Note that if N is the foot of the perpendicular from P to OA (produced if necessary beyond O or beyond A), then $ON=OP|\cos\theta|$.

The work in § 1.2 shows that the orthogonal projections of OP on rectangular cartesian axes with origin O are the x, y, z coordinates of P relative to these axes. We now extend this result to find the orthogonal projection of OP on a line OA which is not necessarily part of one of the coordinate axes.

Let the direction cosines of OA be l, m, n and let P be the point (x, y, z). Then the orthogonal projection of OP on OA is

$$lx + my + nz. \tag{1.11}$$

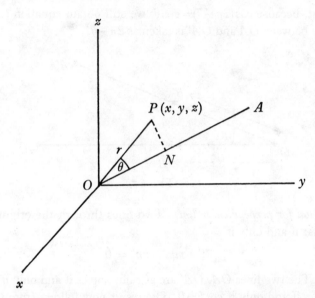

Fig. 7. ON is the orthogonal projection of OP on the line OA

Proof. By equations (1.4), the direction cosines of OP are x/r, y/r, z/r, where $r = OP$. Hence, by formula (1.9), the angle θ between OP and OA is given by

$$\cos \theta = (lx + my + nz)/r.$$

From the definition of the orthogonal projection of OP on OA, expression (1.11) follows at once.

EXERCISES

17. Points A, B have coordinates $(1, 4, -1)$, $(-1, 3, 2)$ respectively. If O is the origin, find the point P on OA produced which is such that the orthogonal projection of OP on OB is of length $9\sqrt{14}/7$.

18. A line OP joins the origin O to the point $P(3, 1, 5)$. Show that the orthogonal projection of OP on the line in the positive octant making equal angles with all three axes is $3\sqrt{3}$.

19. The feet of the perpendiculars from the point $(4, -4, 0)$ to the lines through the origin whose direction cosines are $(\frac{1}{2}\sqrt{2}, \frac{1}{2}\sqrt{2}, 0)$, $(\frac{1}{3}, \frac{2}{3}, \frac{2}{3})$ are denoted by N, N'. Find the lengths of ON, ON', where O is the origin, and explain why one of these lengths is zero.

1.5 Rotation of axes

The transformation matrix and its properties. Consider two sets of right-handed rectangular cartesian axes $Oxyz$, $Ox'y'z'$. It is easily seen that, by a suitable continuous movement about O, the set of axes $Oxyz$ (with Ox, Oy, Oz always fixed relative to each other) may be brought into coincidence with the set $Ox'y'z'$. Such a movement will be called a *rotation* of the axes.

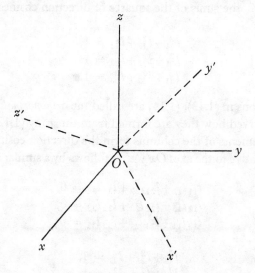

FIG. 8. A rotation of axes

Note that if one set of axes is right-handed and the other left-handed, it is impossible to bring them into coincidence by a rotation. It will be convenient to refer to $Oxyz$ as the original axes and $Ox'y'z'$ as the new axes.

Let the direction cosines of Ox' relative to the axes $Oxyz$ be l_{11}, l_{12}, l_{13}. Further, denote the direction cosines of Oy' and Oz' by l_{21}, l_{22}, l_{23} and l_{31}, l_{32}, l_{33}. We may conveniently summarize this by the following array:

O	x	y	z
x'	l_{11}	l_{12}	l_{13}
y'	l_{21}	l_{22}	l_{23}
z'	l_{31}	l_{32}	l_{33}

(1.12)

In this array, the direction cosines of Ox' relative to the axes $Oxyz$ occur in the first row, the direction cosines of Oy' occur in the second row, and those of Oz' in the third row. Furthermore, reading down the three columns in turn, it is easily seen that we obtain the direction cosines of the axes Ox, Oy, Oz relative to the axes $Ox'y'z'$. The array of direction cosines in (1.12) is called the *transformation matrix*.

Since the axes Ox', Oy', Oz' are mutually perpendicular,

$$l_{11}l_{21}+l_{12}l_{22}+l_{13}l_{23} = 0,$$
$$l_{21}l_{31}+l_{22}l_{32}+l_{23}l_{33} = 0, \qquad (1.13)$$
$$l_{31}l_{11}+l_{32}l_{12}+l_{33}l_{13} = 0.$$

Also, from § 1.2, the sums of the squares of direction cosines are all unity, and so

$$l_{11}^2+l_{12}^2+l_{13}^2 = 1,$$
$$l_{21}^2+l_{22}^2+l_{23}^2 = 1, \qquad (1.14)$$
$$l_{31}^2+l_{32}^2+l_{33}^2 = 1.$$

The six equations in (1.13), (1.14) are called the *orthonormality relations*: it should be observed how they are formed from the array (1.12).

Since the elements of the columns form the direction cosines of the axes Ox, Oy, Oz relative to the axes $Ox'y'z'$, it follows by a similar argument that

$$l_{11}l_{12}+l_{21}l_{22}+l_{31}l_{32} = 0,$$
$$l_{12}l_{13}+l_{22}l_{23}+l_{32}l_{33} = 0, \qquad (1.15)$$
$$l_{13}l_{11}+l_{23}l_{21}+l_{33}l_{31} = 0;$$

and

$$l_{11}^2+l_{21}^2+l_{31}^2 = 1,$$
$$l_{12}^2+l_{22}^2+l_{32}^2 = 1, \qquad (1.16)$$
$$l_{13}^2+l_{23}^2+l_{33}^2 = 1.$$

Equations (1.15) and (1.16) are an important alternative form of the orthonormality relations. They may be derived from equations (1.13) and (1.14) by a purely algebraic argument.

The transformation matrix satisfies one further condition, which arises from the fact that the axes $Oxyz$, $Ox'y'z'$ are both right-handed. Consider the determinant

$$T = \begin{vmatrix} l_{11} & l_{12} & l_{13} \\ l_{21} & l_{22} & l_{23} \\ l_{31} & l_{32} & l_{33} \end{vmatrix}.$$

(For the reader unacquainted with determinants, an account of all the theory needed in this book is included in Appendix 1.)

Denoting the transpose of T by T', we have

$$T^2 = TT' = \begin{vmatrix} l_{11} & l_{12} & l_{13} \\ l_{21} & l_{22} & l_{23} \\ l_{31} & l_{32} & l_{33} \end{vmatrix} \times \begin{vmatrix} l_{11} & l_{21} & l_{31} \\ l_{12} & l_{22} & l_{32} \\ l_{13} & l_{23} & l_{33} \end{vmatrix}.$$

Hence, multiplying the two determinants and using the orthonormality conditions (1.13) and (1.14),

$$T^2 = \begin{vmatrix} 1 & 0 & 0 \\ 0 & 1 & 0 \\ 0 & 0 & 1 \end{vmatrix} = 1.$$

Thus $T = \pm 1$.

Now, when the axes $Oxyz$, $Ox'y'z'$ coincide, it is easily seen that the appropriate values of the direction cosines in the array (1.12) are $l_{ij} = 1$ when $i = j$, $l_{ij} = 0$ when $i \neq j$; and so for this particular case

$$T = \begin{vmatrix} 1 & 0 & 0 \\ 0 & 1 & 0 \\ 0 & 0 & 1 \end{vmatrix} = 1.$$

If the axes are rotated out of coincidence, the direction cosines l_{ij} will vary during the rotation in a continuous manner (i.e. with no 'sudden jumps' in value), and as the determinant T is the sum of products of the direction cosines its value also will vary continuously. But at all stages of the rotation $T = 1$ or -1, and so, for no discontinuity in value to occur, T must take the value 1 *throughout* the rotation; or else the value -1. Since $T - 1$ when the two coordinate systems coincide, it follows that in all positions

$$T = \begin{vmatrix} l_{11} & l_{12} & l_{13} \\ l_{21} & l_{22} & l_{23} \\ l_{31} & l_{32} & l_{33} \end{vmatrix} = 1. \tag{1.17}$$

This is the additional condition to be satisfied by the transformation matrix.

We have shown that, if the components of the array (1.12) are the direction cosines of the new axes relative to the original axes, conditions (1.13), (1.14) and (1.17) are necessarily satisfied. These conditions are also sufficient for the array to represent a rotation of right-handed axes $Oxyz$. For firstly, if equations (1.13) are satisfied the axes Ox', Oy', Oz' are mutually perpendicular; secondly, if equations (1.14) are satisfied the rows in the transformation matrix represent direction cosines of Ox', Oy', Oz'; and finally, if (1.17) is satisfied the system $Ox'y'z'$ is right-handed.

Transformation of coordinates. Let a point P have coordinates (x, y, z) and (x', y', z') relative to the axes $Oxyz$ and $Ox'y'z'$ respectively. The x'-, y'-,

z'-coordinates of P are the orthogonal projections of OP on Ox', Oy', Oz'.
Hence, using (1.11) to calculate these, we obtain

$$\begin{aligned}
x' &= l_{11}x + l_{12}y + l_{13}z, \\
y' &= l_{21}x + l_{22}y + l_{23}z, \\
z' &= l_{31}x + l_{32}y + l_{33}z.
\end{aligned} \tag{1.18}$$

Equations (1.18) show how the coordinates of P transform under a rotation
of axes; it should be noted how these expressions are formed from the
array (1.12).

We could, of course, regard the axes $Ox'y'z'$ as the original set and the
axes $Oxyz$ as the new set, and determine the coordinates (x, y, z) in terms
of (x', y', z'). Remembering that the elements of the columns in (1.12) are
the direction cosines of the x-, y-, z-axes relative to the axes $Ox'y'z'$, it
follows by using (1.11) again that

$$\begin{aligned}
x &= l_{11}x' + l_{21}y' + l_{31}z', \\
y &= l_{12}x' + l_{22}y' + l_{32}z', \\
z &= l_{13}x' + l_{23}y' + l_{33}z'.
\end{aligned} \tag{1.19}$$

The reader may verify as an exercise that equations (1.19) also follow
algebraically from equations (1.18) by solving for x, y, z.

EXERCISES

20. Two sets of axes $Oxyz$, $Ox'y'z'$ are such that the first set may be placed in
the position of the second set by a rotation of 180° about the x-axis. Write down
in the form of array (1.12) the set of direction cosines which corresponds to this
rotation. If a point has coordinates $(1,1,1)$ relative to the axes $Oxyz$, find its
coordinates relative to the axes $Ox'y'z'$.

21. A set of axes $Ox'y'z'$ is initially coincident with a set $Oxyz$. The set $Ox'y'z'$
is then rotated through an angle θ about the z-axis, the direction of rotation being
from the x-axis to the y-axis. Show that

$$\begin{aligned}
x' &= x\cos\theta + y\sin\theta, \\
y' &= -x\sin\theta + y\cos\theta, \\
z' &= z.
\end{aligned}$$

[*Hint.* Consider the direction cosines of the new axes, and use equations (1.18).]

22. Show that the following equations represent a rotation of a set of axes
about a fixed point:

$$\begin{aligned}
x' &= x\sin\theta\cos\phi + y\sin\theta\sin\phi + z\cos\theta, \\
y' &= x\cos\theta\cos\phi + y\cos\theta\sin\phi - z\sin\theta, \\
z' &= -x\sin\phi + y\cos\phi.
\end{aligned}$$

[*Hint.* Show that the coefficients of x, y, z satisfy (1.17) and the orthonormality conditions.]

23. Solve the equations of Exercise 22 for x, y, z in terms of x', y', z'. [*Hint.* Multiply the first equation by $\sin\theta\cos\phi$, the second by $\cos\theta\cos\phi$, the third by $-\sin\phi$, and add to obtain

$$x = x' \sin\theta\cos\phi + y' \cos\theta\cos\phi - z' \sin\phi.$$

Similarly for y and z.]

24. With reference to the transformation array (1.12) in the text, show that

$$l_{11} = l_{22}l_{33} - l_{23}l_{32},$$
$$l_{12} = l_{23}l_{31} - l_{21}l_{33},$$
$$l_{13} = l_{21}l_{32} - l_{22}l_{31}.$$

Write down two sets of three similar relations. [*Hint.* By using (1.17), show that if l_{11}, l_{12}, l_{13} have the values given above then $l_{11}^2 + l_{12}^2 + l_{13}^2 = 1$. The solution to this exercise is given with the answers at the end of the book.]

1.6   The summation convention and its use

It is possible to simplify the statement of equations (1.18) and (1.19) by relabelling the coordinates (x, y, z) as (x_1, x_2, x_3) and the coordinates (x', y', z') as (x_1', x_2', x_3'). With this change of notation, equations (1.18) become

$$x_1' = l_{11}x_1 + l_{12}x_2 + l_{13}x_3 = \sum_{j=1}^{3} l_{1j}x_j \qquad (1.20)$$

and

$$x_2' = \sum_{j=1}^{3} l_{2j}x_j, \qquad (1.21)$$

$$x_3' = \sum_{j=1}^{3} l_{3j}x_j. \qquad (1.22)$$

Even more briefly, we may write equations (1.20) to (1.22) in the form

$$x_i' = \sum_{j=1}^{3} l_{ij}x_j, \quad i = 1, 2, 3. \qquad (1.23)$$

Similarly, equations (1.19) may be reduced to the form

$$x_i = \sum_{j=1}^{3} l_{ji}x_j', \quad i = 1, 2, 3. \qquad (1.24)$$

In equations (1.23), (1.24) the suffix j appears twice in the sums on the right-hand sides, and we sum over all three possible values of j. This situation occurs so frequently that it is convenient to adopt a convention which often avoids the necessity of writing summation signs.

SUMMATION CONVENTION. *Whenever a suffix appears twice in the same expression, that expression is to be summed over all possible values of the suffix.*

Using the summation convention, equations (1.23), (1.24) become simply

$$x'_i = l_{ij} x_j, \tag{1.25}$$
$$x_i = l_{ji} x'_j. \tag{1.26}$$

It is understood here that when a suffix is used alone in an equation (such as i on the left-hand and right-hand sides of (1.25) and (1.26)) the equation under consideration holds for each value of that suffix. The reader should notice that equations (1.25), (1.26) are much more elegant and more convenient than the original forms (1.18), (1.19).

Kronecker delta. The Kronecker delta is defined by

$$\delta_{ij} = \begin{cases} 0 & \text{when } i \neq j \\ 1 & \text{when } i = j. \end{cases} \tag{1.27}$$

By introducing this symbol and using the summation convention, the orthonormality relations (1.13), (1.14) are embodied in the single equation

$$l_{ik} l_{jk} = \delta_{ij}. \tag{1.28}$$

For example, taking $i=1, j=2$, this becomes

$$l_{1k} l_{2k} = 0;$$

that is,

$$l_{11} l_{21} + l_{12} l_{22} + l_{13} l_{23} = 0,$$

which is $(1.13)_1$. Again, taking $i=j=1$ in (1.28),

$$l_{1k} l_{1k} = 1;$$

that is

$$l_{11}^2 + l_{12}^2 + l_{13}^2 = 1,$$

which is $(1.14)_1$. Similarly, taking the other possible combinations of the suffixes i, j we may obtain the remaining four orthonormality conditions.

The alternative form of the orthonormality conditions as expressed by (1.15), (1.16) are embodied in the equation

$$l_{ki} l_{kj} = \delta_{ij}. \tag{1.29}$$

The Kronecker delta is a useful symbol in many contexts other than vector analysis.

Further remarks on the summation convention

(i) A repeated suffix is known as a *dummy suffix*, because it may be replaced by any other suitable symbol. For example,

$$l_{1j} l_{2j} = l_{1k} l_{2k} = l_{1\alpha} l_{2\alpha},$$

since in each expression summation over the repeated suffix is implied.

(ii) When the summation convention is in use, care must be taken to avoid using any suffix more than twice in the same expression. (The meaning of $l_{1j}l_{jj}$, for example, is not clear.)

(iii) As far as we are concerned in this book there are only three possible values for a suffix, namely 1, 2 and 3. The reader will appreciate, though, that elsewhere it might be convenient to increase or decrease the range of a suffix.

Some parts of vector analysis can be shortened considerably by using the convention. *We shall usually warn the reader when the summation convention is in use.*

EXERCISES

25. If

$$a_{11} = 1, \quad a_{12} = -1, \quad a_{13} = 0,$$
$$a_{21} = -2, \quad a_{22} = 3, \quad a_{23} = 1,$$
$$a_{31} = 2, \quad a_{32} = 0, \quad a_{33} = 4,$$

show that

$$a_{ii} = 8, \quad a_{i1}a_{i2} = -7, \quad a_{i2}a_{i3} = 3,$$
$$a_{1i}a_{2i} = -5, \quad a_{2i}a_{3i} = 0, \quad a_{i1}a_{2i} = -6.$$

26. If the numbers a_{ij} are as given in Exercise 25 and if

$$b_1 = 1, \quad b_2 = -1, \quad b_3 = 4,$$

show that

$$a_{1i}b_i = 2, \quad a_{j1}b_j = 11, \quad a_{ji}a_{i1}b_j = 49.$$

[*Hint.* For the last part, evaluate $a_{j1}b_j$, $a_{j2}b_j$, and $a_{j3}b_j$ first.]

27. Show that

$$\delta_{ij}b_j = \delta_{ji}b_j = b_i.$$

28. If the numbers a_{ij} are as given in Exercise 25, evaluate:

(i) $a_{1j}\delta_{1j}$, (ii) $a_{12}\delta_{ii}$, (iii) $a_{1i}a_{2k}\delta_{ik}$.

29. The suffix i may assume all integral values from 0 to ∞. If the numbers a_i, b_i, c_i are defined by

$$a_i = x^i, \quad b_i = \frac{1}{i!}, \quad c_i = (-1)^i,$$

where x is a constant and (by definition) $0! = 1$, show that

$$a_i b_i = e^x, \quad a_i c_i = \frac{1}{1+x}.$$

30. If the quantities e_{ij}, e'_{ij} satisfy the relation

$$l_{mi}l_{nj}e_{ij} = e'_{mn},$$

and if
$$l_{ki}l_{kj} = \delta_{ij},$$
show that
$$e_{ij} = l_{mi}l_{nj}e'_{mn}.$$

[*Hint.* Multiply the first equation by $l_{mp}l_{nq}$.]

1.7   Invariance with respect to a rotation of the axes

Consider a function $f(\alpha_1, \alpha_2, \dots)$ of several elements α_1, α_2, …, such that given any set of rectangular cartesian axes $Oxyz$, the elements α_1, α_2, … are determined by a definite rule. Denote by α'_1, α'_2, … the elements corresponding to any other set of rectangular cartesian axes $Ox'y'z'$, with the same origin O. Then, if

$$f(\alpha'_1, \alpha'_2, \dots) = f(\alpha_1, \alpha_2, \dots),$$

the function f is said to be *invariant with respect to a rotation of the axes.*

The examples which follow should clarify the idea of invariance.

Examples of invariants

(1) The function $\sqrt{(x^2+y^2+z^2)}$ is invariant, since (1.19) gives

$$\sqrt{(x^2+y^2+z^2)} = \sqrt{\{(l_{11}^2+l_{12}^2+l_{13}^2)\,x'^2 + (l_{21}^2+l_{22}^2+l_{23}^2)\,y'^2}$$
$$+ (l_{31}^2+l_{32}^2+l_{33}^2)\,z'^2 + 2(l_{11}\,l_{21}+l_{12}\,l_{22}+l_{13}\,l_{23})\,x'\,y'$$
$$+ 2(l_{21}\,l_{31}+l_{22}\,l_{32}+l_{23}\,l_{33})\,y'\,z'$$
$$+ 2(l_{31}\,l_{11}+l_{32}\,l_{12}+l_{33}\,l_{13})\,z'\,x'\}.$$

Using the orthonormality conditions (1.13) and (1.14), this reduces to

$$\sqrt{(x^2+y^2+z^2)} = \sqrt{(x'^2+y'^2+z'^2)}.$$

This result has an immediate geometrical interpretation: it expresses the fact that the distance between the origin O and a point P does not depend upon the system of coordinates used in calculating the distance.

The proof given above can be shortened considerably by using the summation convention (see Exercise 32 at the end of this section).

(2) If OA and OB are two lines through the origin, the expression representing the cosine of the angle between them is clearly invariant with respect to a rotation of the axes. To verify this algebraically, let (a_1, a_2, a_3), (b_1, b_2, b_3) be the coordinates of A, B relative to the axes $Oxyz$. If $OA=a$, $OB=b$, the direction cosines of OA and OB are

$$\frac{a_1}{a}, \frac{a_2}{a}, \frac{a_3}{a} \quad \text{and} \quad \frac{b_1}{b}, \frac{b_2}{b}, \frac{b_3}{b}.$$

If θ is the angle between OA, OB, formula (1.9) gives

$$\cos\theta = \frac{a_1 b_1 + a_2 b_2 + a_3 b_3}{ab} = \frac{a_i b_i}{ab},$$

using the summation convention. From Example (1) we see that a and b are invariant with respect to a rotation of the axes. Thus, to show that $\cos\theta$ is invariant it only remains to show that $a_i b_i$ is invariant.

Using (1.25) we see that on transforming to new axes $Ox'y'z'$, the coordinates of A and B become (a_1', a_2', a_3') and (b_1', b_2', b_3'), where

$$a_i' = l_{ij} a_j, \quad b_i' = l_{ik} b_k.$$

Thus,

$$a_i' b_i' = l_{ij} l_{ik} a_j b_k. \tag{1.30}$$

(Notice that before forming the expression for $a_i' b_i'$, different dummy suffixes must be used in the formulae for a_i' and b_i'; otherwise a suffix would appear more than twice in the right-hand side of (1.30).) By using the orthonormality relations in the form (1.29) (with the suffixes changed to those required here), (1.30) becomes

$$\begin{aligned} a_i' b_i' &= \delta_{jk} a_j b_k \\ &= a_k b_k = a_i b_i. \end{aligned}$$

It follows that the quantity $a_i b_i (= a_1 b_1 + a_2 b_2 + a_3 b_3)$ is invariant under a rotation of the axes, as required, and that $\cos\theta$ is invariant also.

The concept of invariance with respect to rotation of the axes is important because the recognizable aspects of a physical system are usually invariant in this way.

For example, the distance between two points, the volume of a specified region, and the resolute of a force along a given line are all independent of any special coordinate system, and the expressions which represent them are invariant with respect to a rotation of the axes.

EXERCISES

31. Find the x', y', z' coordinates of the points $x=1$, $y=1$, $z=0$ and $x=0$, $y=1$, $z=1$ for the rotation of axes given in Exercise 21, namely

$$\begin{aligned} x' &= x\cos\theta + y\sin\theta, \\ y' &= -x\sin\theta + y\cos\theta, \\ z' &= z. \end{aligned}$$

Verify that the angle between the lines joining the origin to these two points works out as 60° with either set of axes.

32. Show that the quantity $x_1^2 + x_2^2 + x_3^2 = x_i x_i$ is invariant under a rotation of axes.

1.8 Matrix notation

Another way of expressing some of the results obtained in this chapter is afforded by the use of matrices. It is not our intention to make much use of matrix notation in this book, but those readers familiar with matrices may welcome the following brief remarks.

The matrix of direction cosines in (1.12) may be denoted by

$$L = \begin{pmatrix} l_{11} & l_{12} & l_{13} \\ l_{21} & l_{22} & l_{23} \\ l_{31} & l_{32} & l_{33} \end{pmatrix}; \tag{1.31}$$

its transpose is

$$L^T = \begin{pmatrix} l_{11} & l_{21} & l_{31} \\ l_{12} & l_{22} & l_{32} \\ l_{13} & l_{23} & l_{33} \end{pmatrix}. \tag{1.32}$$

With this notation, the orthonormality conditions (1.13) and (1.14) may be expressed as

$$LL^T = I, \tag{1.33}$$

where

$$I = \begin{pmatrix} 1 & 0 & 0 \\ 0 & 1 & 0 \\ 0 & 0 & 1 \end{pmatrix} \tag{1.34}$$

is the unit matrix. By pre-multiplying (1.33) by L^{-1}, which is the inverse of L, the result that

$$L^T = L^{-1}. \tag{1.35}$$

is obtained.

Writing

$$\mathbf{x} = \begin{pmatrix} x_1 \\ x_2 \\ x_3 \end{pmatrix} \quad \text{and} \quad \mathbf{x}' = \begin{pmatrix} x_1' \\ x_2' \\ x_3' \end{pmatrix}, \tag{1.36}$$

the transformation rules (1.25) and (1.26) may be expressed as

$$\mathbf{x}' = L\mathbf{x} \tag{1.37}$$

and

$$\mathbf{x} = L^T\mathbf{x}', \tag{1.38}$$

respectively. Expression (1.38) may be derived at once from (1.37) by pre-multiplying by L^{-1} and using (1.35).

CHAPTER 2

SCALAR AND VECTOR ALGEBRA

2.1 Scalars

Any mathematical entity or any property of a physical system which can be represented by a single real number is called a *scalar*. In the case of a physical property, the single real number will be a measure of the property in some chosen units (e.g. kilogrammes, metres, seconds).

Particular examples of scalars are: (i) the modulus of a complex number, (ii) mass, (iii) volume, (iv) temperature. Note that real numbers are themselves scalars.

Single letters printed in italics (such as a, b, c, etc.) will be used to denote real numbers representing scalars. For convenience, statements like 'let a be a real number representing a scalar' will be abbreviated to 'let a be a scalar'.

Equality of scalars. Two scalars (measured in the same units if they are physical properties) are said to be equal if the real numbers representing them are equal.

It will be assumed throughout this book that in the case of physical entities the same units are used on both sides of any equality sign.

Scalar addition, subtraction, multiplication, and division. The sum of two scalars is defined as the sum of the real numbers representing them; and similarly, scalar subtraction, multiplication, and division are defined by the corresponding operations on the representative numbers. In the case of physical scalars, the operations of addition and subtraction are physically meaningful only for similar scalars, such as two lengths or two masses.

Some care is necessary in the matter of units. For example, if a, b are two physical scalars, it is meaningful to say their sum is $a+b$ only if the units of measurement are the same.

Again, consider the equation $T = \frac{1}{2}mv^2$ giving the kinetic energy T of a particle of mass m travelling with speed v. If T has the value 30 kg m^2/s^2 and v has the value 0·1 km/s, then to calculate $m = 2T/v^2$, consistent units for length and time must first be introduced. Thus, converting the given speed to m/s, we find v has the value 100 m/s. Hence the value of m is $2 \times 30/10000 = 0·006$ kg.

Henceforth it is to be understood that consistent units of measurement are used in operations involving physical properties.

2.2   Vectors: basic notions

From an elementary standpoint, the reader will probably have already encountered properties of physical systems which require for their complete specification a scalar magnitude and a direction – the velocity of a moving point and the force on a body are particular examples – and such properties are called vectors. We define a vector formally below; but as the definition might otherwise seem strange, we shall try to put it into perspective by continuing for the moment on an intuitive basis.

Consider the velocity of a point P moving relative to fixed rectangular cartesian coordinate axes $Oxyz$. Denote by v_1, v_2, v_3 the rate at which P is travelling away from the yz-, zx-, xy-planes in the directions x-, y-, z-increasing, respectively. Then v_1, v_2, v_3 are called the components of the velocity; together, they describe completely the instantaneous motion of P.

If the axes $Oxyz$ are moved to new fixed positions $O'x'y'z'$ without rotation, the velocity components of P relative to the new axes will be v_1, v_2, v_3, as before: for, v_1 is the rate at which P moves away from the yz-plane and this will be the same as the rate at which it moves away from the $y'z'$-plane, since the two planes are parallel; and similarly for the other two directions.

If the axes are rotated about O to new positions $Ox'y'z'$, the velocity components, v_1', v_2', v_3' say, of P relative to the new axes must be related in some way to the original components; for the two sets of components are each sufficient to define the motion of P. The relationship will depend upon the relative position of the two sets of axes (which is defined by the direction cosines of one set relative to the other), and by continuing the discussion on an intuitive basis it could be obtained explicitly.

The formal definition which follows will be seen to fit into the pattern which the brief remarks above suggest. That vectors do indeed have an associated magnitude and direction will follow later in this section as a consequence of the definition, and so the consistency of this with intuitive ideas will be confirmed.

DEFINITION. A *vector* is any mathematical or physical entity which is such that:

(i) when it is associated with a set of rectangular cartesian axes $Oxyz$ it can be represented completely by three scalars a_1, a_2, a_3 related, in turn, to the axes of x, y, and z;

(ii) the triad of scalars in (i) is invariant under a translation of the axes;

i.e. if $Oxyz$, $O'x'y'z'$ are rectangular cartesian axes (with different origins O, O') such that Ox is parallel to $O'x'$, Oy is parallel to $O'y'$ and Oz is parallel to $O'z'$, and if the triads associated with the two coordinate systems are (a_1, a_2, a_3), (a'_1, a'_2, a'_3) repectively, then $a_1 = a'_1$, $a_2 = a'_2$, $a_3 = a'_3$;

(iii) if the triads of scalars associated with two sets of axes $Oxyz$, $Ox'y'z'$ (with the same origin O) are (a_1, a_2, a_3), (a'_1, a'_2, a'_3) respectively, and if the direction cosines of Ox', Oy', Oz' relative to the axes $Oxyz$ are given by the transformation matrix (1.12), then

$$\begin{aligned}
a'_1 &= l_{11}a_1 + l_{12}a_2 + l_{13}a_3, \\
a'_2 &= l_{21}a_1 + l_{22}a_2 + l_{23}a_3, \\
a'_3 &= l_{31}a_1 + l_{32}a_2 + l_{33}a_3.
\end{aligned} \tag{2.1}$$

Introducing the summation convention, equations (2.1) reduce to

$$a'_i = l_{ij}a_j \quad (i = 1, 2, 3). \tag{2.2}$$

Vectors will be denoted by letters printed in heavy type. Thus

$$\mathbf{a} = (a_1, a_2, a_3); \tag{2.3}$$

and a_1, a_2, a_3 are called respectively the x-component, y-component, z-component of the vector $\mathbf{a}$.

To avoid repetition, components of all vectors are henceforth to be taken as referred to axes $Oxyz$, unless stated otherwise.

Equality of vectors. Two vectors $\mathbf{a} = (a_1, a_2, a_3)$ and $\mathbf{b} = (b_1, b_2, b_3)$, referred to the same coordinate system, are defined to be equal if and only if their respective components are equal; i.e.

$$\mathbf{a} = \mathbf{b} \Leftrightarrow a_1 = b_1, a_2 = b_2, a_3 = b_3. \tag{2.4}$$

The zero vector. The vector whose components are all zero is called the zero vector or null vector, and is written as

$$\mathbf{0} = (0, 0, 0). \tag{2.5}$$

The position vector. Let A and B be points whose coordinates relative to axes $Oxyz$ are (a_1, a_2, a_3) and (b_1, b_2, b_3), respectively. The position of B relative to A is a vector written

$$\overrightarrow{AB} = (b_1 - a_1, \ b_2 - a_2, \ b_3 - a_3);$$

it is called the *position vector of B relative to A*.

Proof. To prove that $\overrightarrow{AB}$ is a vector it is necessary to establish that conditions (i) to (iii) of the definition are satisfied.

(i) The scalars b_1-a_1, b_2-a_2, b_3-a_3 are the coordinates of B relative to axes $Ax'y'z'$ drawn through A and parallel to the original axes $Oxyz$ (Fig. 9). These scalars are clearly related to the axes Ox, Oy, Oz, respectively, and they define completely the position of B relative to A (within the coordinate system $Oxyz$).

(ii) Suppose that the axes $Oxyz$ are moved parallel to themselves so that they pass through a new origin whose coordinates are $(-x_0, -y_0, -z_0)$ referred to the axes in their original position. The coordinates of A become

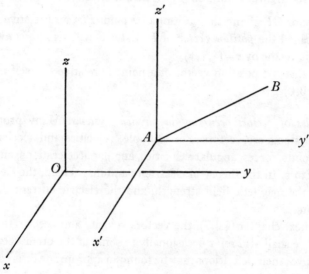

FIG. 9

$(x_0+a_1, y_0+a_2, z_0+a_3)$ and the coordinates of B become $(x_0+b_1, y_0+b_2, z_0+b_3)$. Thus, if $(\overrightarrow{AB})'$ denotes the position vector referred to the new axes,

$$(\overrightarrow{AB})' = [(x_0+b_1)-(x_0+a_1), (y_0+b_2)-(y_0+a_2), (z_0+b_3)-(z_0+a_3)]$$
$$= (b_1-a_1, b_2-a_2, b_3-a_3).$$

It follows that the components of $\overrightarrow{AB}$ are invariant with respect to a translation of the axes.

(iii) Let Ox', Oy', Oz' be rectangular coordinate axes whose direction cosines relative to the axes $Oxyz$ are given by (1.12). Using (1.23), the coordinates (a_1', a_2', a_3') and (b_1', b_2', b_3') of A and B referred to the axes $Ox'y'z'$ are

$$a_i' = \sum_{j=1}^{3} l_{ij} a_j, \quad b_i' = \sum_{j=1}^{3} l_{ij} b_j \quad (i = 1, 2, 3),$$

giving

$$b_i' - a_i' = \sum_{j=1}^{3} l_{ij}(b_j - a_j) \quad (i = 1, 2, 3). \tag{2.6}$$

It follows that the components of $\overrightarrow{AB}$ obey the vector transformation law (2.2). As all three conditions of the definition are satisfied, $\overrightarrow{AB}$ is a vector.

The vector $\overrightarrow{OP}$ giving the position of a point $P(x, y, z)$ relative to an origin O is called the *position vector* of P. Later it will also be convenient to denote this vector by $\mathbf{r} = (x, y, z)$.

Notice that the position vector of a point O relative to itself is the zero vector $(0, 0, 0)$.

Examples of vectors occurring in physical systems. Many properties of physical systems are vectors. For example, velocities and accelerations of moving points, forces, angular velocities, angular accelerations, and couples are all vectors. In the theory of electricity and magnetism, the electric field strength, the magnetic field strength, and the electric current density are also vectors.

In the next chapter (§ 3.7), the vectors velocity and acceleration will be defined. We shall also refer occasionally to some of the other vectors mentioned above; their definitions may be found in appropriate reference books.

Geometrical representation of vectors. Let $\mathbf{a} = (a_1, a_2, a_3)$ be any non-zero vector, and let A be the point whose x-, y-, z-coordinates are a_1, a_2, a_3. Then $\overrightarrow{OA} = (a_1, a_2, a_3)$, showing that $\mathbf{a}$ and $\overrightarrow{OA}$ have the same components. Thus, the *directed straight line segment OA* may be taken as a geometrical representation of $\mathbf{a}$. When a directed line segment such as OA represents a vector, this may be shown in a diagram by drawing an arrow on the line pointing from O to A (Fig. 10). Any zero vector is appropriately represented geometrically by a single point.

Let $O'A'$ be a line segment equal in length to OA and drawn parallel to and in the same sense as OA. If $O'x'$, $O'y'$, $O'z'$ are axes through O' such that they are parallel to Ox, Oy, Oz, respectively, the coordinates of A' referred to these new axes will be (a_1, a_2, a_3). Hence $\overrightarrow{O'A'}$ is a second geometrical representation of the vector $\mathbf{a}$. It follows that the geometrical representation of **a** as a directed line segment is not unique: any other directed line segment $O'A'$ which is parallel to OA and of equal length may also be used to represent **a**.

Direction of a vector. Because a non-zero vector can be represented as a directed line segment, a vector is said to have (or to be associated with) a *direction*: naturally this is taken to be the same as the direction of the directed line segments which represent the vector. Thus, the direction of

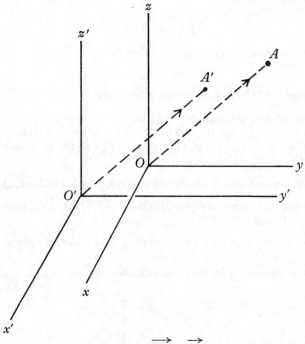

FIG. 10. $\overrightarrow{O'A'} = \overrightarrow{OA}$

the vector **a** is the direction in which OA points. The direction of the null vector is not defined.

Two vectors are said to be *parallel* if they are in the same direction, and *anti-parallel* if they are in opposite directions.

Magnitude of a vector. The *magnitude* of a vector $\mathbf{a} = (a_1, a_2, a_3)$ is defined as

$$a = \sqrt{(a_1^2 + a_2^2 + a_3^2)}. \tag{2.7}$$

It will also be convenient occasionally to denote the magnitude of a vector **a** by $|\mathbf{a}|$.

If **a** is represented geometrically as $\overrightarrow{OA} = (a_1, a_2, a_3)$, it is seen that a is just the length of the line segment OA. Since this length is invariant under a rotation of the axes (cf. § 1.7, Example 1), it follows that the magnitude of a vector is also invariant. The magnitude of a vector is also sometimes termed its *modulus* or its *norm*.

Unit vectors. A vector of unit magnitude is called a *unit vector*. Unit vectors are frequently distinguished by a circumflex; thus $\hat{\mathbf{r}} = (\cos\theta, \sin\theta, 0)$ is a unit vector.

If **a** is any vector, the unit vector whose direction is that of **a** is denoted by **â**.

EXERCISES

1. Show that equations (2.2) are equivalent to

$$a_k = l_{ik}a_i'.$$

[*Hint.* Multiply (2.2) by l_{ik} and use the orthonormality conditions (1.29).]

2. Relative to axes $Oxyz$, points P, Q have coordinates $(1,2,3)$, $(0,0,1)$. Find the components, referred to the axes $Oxyz$, of: (i) the position vector of P relative to O; (ii) the position vector of O relative to P; (iii) the position vector of P relative to Q.

3. Find the magnitudes of the vectors $\mathbf{a} = (1, 3, 4)$ and $\mathbf{b} = (2, -1, 0)$.

4. Show that the vectors $\mathbf{a} = (0, -3, 3)$, $\mathbf{b} = (0, -5, 5)$ have the same direction. What is the ratio a/b?

5. The transformation matrix for a rotation from axes $Oxyz$ to axes $Ox'y'z'$ is as follows:

O	x	y	z
x'	0	1	0
y'	-1	0	0
z'	0	0	1

Describe (in words or by a diagram) how the positions of the two sets of axes are related.

Relative to the set of axes $Oxyz$, a vector **a** has components $(2,1,2)$. Find the components of **a** relative to the axes $Ox'y'z'$.

6. Show that $\mathbf{a} = (\cos\theta, \sin\theta\cos\phi, \sin\theta\sin\phi)$ is a unit vector.

7. Find two unit vectors which are perpendicular to each of the vectors $\mathbf{a} = (0, 0, 1)$, $\mathbf{b} = (0, 1, 1)$. Are there any more unit vectors which are perpendicular to both **a** and **b**?

2.3 Multiplication of a vector by a scalar

If the components of a vector **a** in every rectangular cartesian coordinate system are each multiplied by a scalar λ, we define the ordered **triads** of scalars so formed to be components of $\lambda\mathbf{a}$ (or $\mathbf{a}\lambda$). Thus, if $\mathbf{a}=(a_1, a_2, a_3)$ when referred to axes $Oxyz$,

$$\lambda\mathbf{a} = \mathbf{a}\lambda = (\lambda a_1, \lambda a_2, \lambda a_3). \tag{2.8}$$

It is easily seen that $\lambda\mathbf{a}$ is a vector. For, first, condition (i) of the definition given in § 2.2 is obviously satisfied; and so also is condition (ii), since the triad (a_1, a_2, a_3) is invariant under a translation of the axes. Further, referred to the axes $Ox'y'z'$ (defined in § 2.2),

$$\mathbf{a} = (a_1', a_2', a_3'),$$

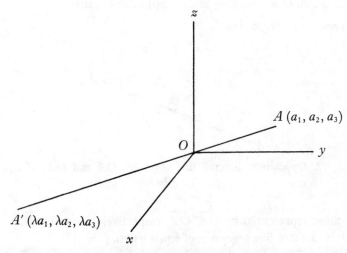

FIG. 11. The case $\lambda < 0$

and hence the components of $\lambda\mathbf{a}$ referred to these axes are $(\lambda a_1', \lambda a_2', \lambda a_3')$. To satisfy (iii) we require

$$\lambda a_i' = l_{ij}(\lambda a_j), \quad (i = 1, 2, 3)$$

and as these are the same as equations (2.2) multiplied by λ, the result follows. Note that when $\lambda = 0$, $\lambda\mathbf{a}$ is the null vector defined by equation (2.5).

From equation (2.8) and the definition of the magnitude of a vector,

$$\begin{aligned}
|\lambda\mathbf{a}| &= \sqrt{(\lambda^2 a_1^2 + \lambda^2 a_2^2 + \lambda^2 a_3^2)} \\
&= |\lambda|\sqrt{(a_1^2 + a_2^2 + a_3^2)} \\
&= |\lambda|a.
\end{aligned}$$

Thus, when a vector is multiplied by a scalar λ, its magnitude is multiplied by $|\lambda|$.

Multiplication of a non-zero vector by a positive scalar leaves the direction of the vector unchanged, but multiplication by a negative scalar reverses its direction. To see this, let $\overrightarrow{OA} = (a_1, a_2, a_3)$ represent $\mathbf{a}$; $\lambda\mathbf{a}$ will then be represented by $\lambda\overrightarrow{OA} = (\lambda a_1, \lambda a_2, \lambda a_3) = \overrightarrow{OA'}$, say. Since A, A' have coordinates (a_1, a_2, a_3), $(\lambda a_1, \lambda a_2, \lambda a_3)$, respectively, the direction cosines of OA and OA' will be

$$\frac{a_1}{a}, \frac{a_2}{a}, \frac{a_3}{a} \quad \text{and} \quad \frac{\lambda a_1}{|\lambda|a}, \frac{\lambda a_2}{|\lambda|a}, \frac{\lambda a_3}{|\lambda|a},$$

where $a = \sqrt{(a_1^2 + a_2^2 + a_3^2)}$. If $\lambda > 0$, $\lambda/|\lambda| = 1$, whilst if $\lambda < 0$, $\lambda/|\lambda| = -1$. It follows that OA, OA' are in the same direction if $\lambda > 0$ and in opposite directions if $\lambda < 0$ (Fig. 11). The direction of $\lambda\mathbf{a}$ is therefore the same as or opposite to that of $\mathbf{a}$ according as λ is positive or negative.

The vector $-\mathbf{a}$. We define

$$-\mathbf{a} = (-1)\mathbf{a}. \tag{2.9}$$

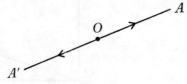

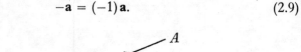

FIG. 12. Oppositely directed line segments OA and OA' of equal lengths

If $\mathbf{a}$, $-\mathbf{a}$ are represented by $\overrightarrow{OA}$, $\overrightarrow{OA'}$ respectively, then OA, OA' will be oppositely directed line segments of equal length (Fig. 12).

Note that $\overrightarrow{OA} = -\overrightarrow{AO}$, since AO and OA', being parallel directed line segments of equal length, represent the same vector.

EXERCISES

8. A point P on a line AB is such that $|AP| : |PB| = 3:2$. Show that

$$2\overrightarrow{AP} = \pm 3\overrightarrow{PB}.$$

9. Show that, for any vector $\mathbf{a}$,

$$\mathbf{a} = a\hat{\mathbf{a}},$$

where $\hat{\mathbf{a}}$ is a unit vector in the same direction as $\mathbf{a}$.

10. Show that the four points with position vectors

$$\mathbf{r}_1, \quad \mathbf{r}_2, \quad \frac{r_2}{r_1}\mathbf{r}_1, \quad \frac{r_1}{r_2}\mathbf{r}_2,$$

where $r_1 \neq 0$, $r_2 \neq 0$, lie on a circle.

2.4 Addition and subtraction of vectors

Addition. The *sum* of two vectors $\mathbf{a} = (a_1, a_2, a_3)$ and $\mathbf{b} = (b_1, b_2, b_3)$ is defined as

$$\mathbf{a} + \mathbf{b} = (a_1 + b_1, a_2 + b_2, a_3 + b_3). \tag{2.10}$$

The set of all vectors is *closed* under the operation of addition; i.e. the sum of two vectors $\mathbf{a}$, $\mathbf{b}$ is a vector. For, considering the requirements (i)–(iii) of the definition of a vector (§ 2.2), it is seen at once that (i) and (ii) are satisfied. Also, if (a_1', a_2', a_3'), (b_1', b_2', b_3') are the components of $\mathbf{a}$, $\mathbf{b}$ referred to the axes $Ox'y'z'$, equations (2.2) give

$$a_i' = l_{ij} a_j \quad \text{and} \quad b_i' = l_{ij} b_j \quad (i = 1, 2, 3).$$

Adding, we have

$$a_i' + b_i' = l_{ij}(a_j + b_j),$$

showing that the components of $\mathbf{a} + \mathbf{b}$ satisfy requirement (iii).

As

$$a_i + b_i = b_i + a_i \quad (i = 1, 2, 3),$$

it follows from (2.10) that

$$\mathbf{a} + \mathbf{b} = \mathbf{b} + \mathbf{a}; \tag{2.11}$$

i.e. the operation of addition is *commutative*.

Also, if $\mathbf{c} = (c_1, c_2, c_3)$ is a third vector, then since

$$(a_i + b_i) + c_i = a_i + (b_i + c_i) \quad (i = 1, 2, 3)$$

we have

$$(\mathbf{a} + \mathbf{b}) + \mathbf{c} = \mathbf{a} + (\mathbf{b} + \mathbf{c}); \tag{2.12}$$

i.e. the operation of addition is *associative*.

It is important to appreciate that rules like the commutative and associative laws of addition cannot be taken for granted by analogy with similar laws for real numbers. A new mathematical system is being established, involving entities which are *not* real numbers; and there is no reason to suppose, until this has been proved, that such rules of manipulation are obeyed.

The triangle law of addition. From the definition of addition of vectors, we may deduce the triangle law of addition (also sometimes called the parallelogram law) as follows:

If two vectors are represented geometrically by $\overrightarrow{AB}$ and $\overrightarrow{BC}$, then their sum is represented by $\overrightarrow{AC}$ (Fig. 13). Thus

$$\overrightarrow{AB} + \overrightarrow{BC} = \overrightarrow{AC}.$$

$\overrightarrow{AC}$ is often called the *resultant* of $\overrightarrow{AB}$ and $\overrightarrow{BC}$.

Proof. Construct axes $Axyz$ through the point A, and draw the line AD so as to complete the parallelogram $ABCD$ (Fig. 14). Let the coordinates of B, D be (b_1, b_2, b_3), (d_1, d_2, d_3) respectively. Then, as BC is parallel to AD,

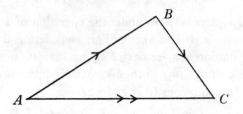

FIG. 13. $\overrightarrow{AB} + \overrightarrow{BC} = \overrightarrow{AC}$ (the triangle law of addition)

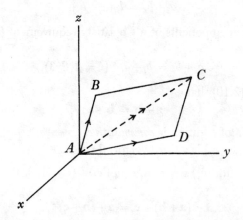

FIG. 14. If $ABCD$ is a parallelogram, $\overrightarrow{AB} + \overrightarrow{AD} = \overrightarrow{AC}$ (the parallelogram law)

the coordinates of C will be $(b_1 + d_1, b_2 + d_2, b_3 + d_3)$. Thus, referred to the axes $Axyz$,

$$\overrightarrow{AB} = (b_1, b_2, b_3),$$
$$\overrightarrow{BC} = \overrightarrow{AD} = (d_1, d_2, d_3),$$
$$\overrightarrow{AC} = (b_1 + d_1, b_2 + d_2, b_3 + d_3).$$

It follows at once from the law of addition of vectors that, as required,

$$\overrightarrow{AB} + \overrightarrow{BC} = \overrightarrow{AC}.$$

Subtraction. The *difference* of two vectors **a**, **b** is defined as

$$\mathbf{a} - \mathbf{b} = \mathbf{a} + (-\mathbf{b}). \tag{2.13}$$

Note that for any vector **a**,

$$\mathbf{a} - \mathbf{a} = \mathbf{0}.$$

Figure 15 shows how the sum and difference of vectors **a**, **b** may be represented geometrically.

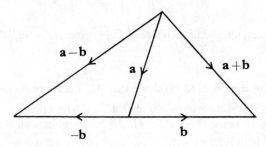

FIG. 15. The sum and difference of vectors **a** and **b**

EXAMPLE 1. The position vectors of three points, A, B, P, relative to an origin O, are such that

$$\overrightarrow{OP} = \frac{\lambda \overrightarrow{OA} + \mu \overrightarrow{OB}}{\lambda + \mu},$$

where λ, μ are non-zero real numbers. Prove that P lies on AB (possibly produced) and that $AP:PB = |\mu| : |\lambda|$ (Fig. 16).

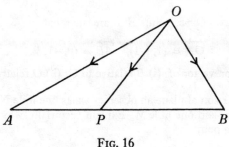

FIG. 16

Solution. By the triangle law,

$$\overrightarrow{OA} = \overrightarrow{OP} + \overrightarrow{PA}$$

and

$$\overrightarrow{OB} = \overrightarrow{OP} + \overrightarrow{PB}.$$

To satisfy the given relation between $\overrightarrow{OP}$, $\overrightarrow{OA}$, $\overrightarrow{OB}$ we require

$$\overrightarrow{OP} = \frac{\lambda(\overrightarrow{OP} + \overrightarrow{PA}) + \mu(\overrightarrow{OP} + \overrightarrow{PB})}{\lambda + \mu},$$

and hence

$$0 = \lambda \overrightarrow{PA} + \mu \overrightarrow{PB}.$$

Since $-\overrightarrow{PA} = \overrightarrow{AP}$, this may be expressed as

$$\overrightarrow{AP} = \frac{\mu}{\lambda} \overrightarrow{PB} \qquad (\lambda, \mu \neq 0). \qquad (2.14)$$

Equating the magnitudes of each side of (2.14) gives $AP = |\mu/\lambda| PB$. Thus

$$AP:PB = |\mu| : |\lambda|.$$

If λ, μ have the same sign, (2.14) shows that $\overrightarrow{AP}$, $\overrightarrow{PB}$ are in the same direction, and so P lies on AB and between A and B. If λ, μ have opposite signs, $\overrightarrow{AP}$, $\overrightarrow{PB}$ are in opposite directions; hence P lies on AB produced beyond B (if $|\mu|/|\lambda| > 1$) or beyond A (if $|\mu|/|\lambda| < 1$).

EXERCISES

11. If **a** and **b** are vectors as given below, verify that their sums and differences are as shown:

	a	b	a+b	a−b
(i)	(2, 2, 2)	(1, 0, 1)	(3, 2, 3)	(1, 2, 1)
(ii)	(3, 0, 0)	(5, 0, 0)	(8, 0, 0)	(−2, 0, 0)
(iii)	(1, −2, 6)	(−1, −3, 7)	(0, −5, 13)	(2, 1, −1).

12. Relative to axes $Oxyz$, points A, B are such that

$$\overrightarrow{OA} = (1, 1, 1), \qquad \overrightarrow{AB} = (0, -1, 3).$$

What is the position vector of: (i) B relative to O; (ii) O relative to B?

13. On a flat horizontal plane an observer walks one mile N, followed by one mile E, one mile S, and one mile W. Explain vectorially why he finds himself back at the starting point.

14. If the angle between vectors **a** and **b** is 60°, and if $a = b = 3$, show that

$$|\mathbf{a} - \mathbf{b}| = 3.$$

15. From the property $AC \leqslant AB + BC$ of a triangle ABC, prove that

$$|\mathbf{a} + \mathbf{b}| \leqslant a + b.$$

For what particular cases is it true that

$$|\mathbf{a} + \mathbf{b}| = a + b?$$

16. Prove that

$$|\mathbf{a} - \mathbf{b}| \leqslant a + b.$$

17. If $\hat{\mathbf{u}}$, $\hat{\mathbf{v}}$ are unit vectors with different directions, show that the vector $\hat{\mathbf{u}} + \hat{\mathbf{v}}$ bisects internally the angle between $\hat{\mathbf{u}}$ and $\hat{\mathbf{v}}$. Is $\frac{1}{2}(\hat{\mathbf{u}} + \hat{\mathbf{v}})$ a unit vector?

2.5   The unit vectors i, j, k

Let $\mathbf{i}, \mathbf{j}, \mathbf{k}$ denote unit vectors in the directions of the x-axis, y-axis, z-axis respectively. Then,

$$\mathbf{i} = (1, 0, 0), \quad \mathbf{j} = (0, 1, 0), \quad \mathbf{k} = (0, 0, 1).$$

Using the rules for multiplication of a vector by a scalar and for addition of vectors, a vector $\mathbf{a} = (a_1, a_2, a_3)$ may be written

$$\mathbf{a} = a_1 \mathbf{i} + a_2 \mathbf{j} + a_3 \mathbf{k}.$$

It is easily shown that, when the triad $\mathbf{i}, \mathbf{j}, \mathbf{k}$ is given, this representation of $\mathbf{a}$ is unique.

The three vectors $\mathbf{i}, \mathbf{j}, \mathbf{k}$ are unit vectors which are mutually perpendicular. Any set of three mutually perpendicular unit vectors is said to be an *orthonormal* set. Because any arbitrary vector can be represented as a linear combination of $\mathbf{i}, \mathbf{j}, \mathbf{k}$, these vectors are also said to form an *orthonormal basis* for the totality of all vectors. Orthonormal bases play an important role in vector analysis.

EXERCISES

18. A unit vector $\mathbf{a}$ in the positive quadrant of the xy-plane makes an angle of $45°$ with each of the axes Ox and Oy. Show that

$$\mathbf{a} = (\mathbf{i} + \mathbf{j})/\sqrt{2}.$$

19. The position vectors of points A and B relative to the origin O of axes $Oxyz$ are $\mathbf{i} - \mathbf{j} + 2\mathbf{k}$ and $5\mathbf{i} + \mathbf{j} + 6\mathbf{k}$ respectively. Show that $AB = 6$.

20. Find a, b, c if

$$(a + b - 2)\mathbf{i} + (c - 1)\mathbf{j} + (a + c)\mathbf{k} = \mathbf{0}.$$

2.6   Scalar products

The *scalar product* (or *dot product*) of two vectors $\mathbf{a} = (a_1, a_2, a_3)$ and $\mathbf{b} = (b_1, b_2, b_3)$ is defined as

$$\mathbf{a} \cdot \mathbf{b} = a_1 b_1 + a_2 b_2 + a_3 b_3. \tag{2.15}$$

This operation between two vectors is commutative, because

$$\begin{aligned} \mathbf{b} \cdot \mathbf{a} &= b_1 a_1 + b_2 a_2 + b_3 a_3 \\ &= \mathbf{a} \cdot \mathbf{b}. \end{aligned} \tag{2.16}$$

The scalar product of **a** with itself is

$$\mathbf{a}.\mathbf{a} = a_1^2 + a_2^2 + a_3^2 = a^2;$$

thus **a**.**a** is the square of the magnitude of **a**.

If $\mathbf{a} = (a_1, a_2, a_3)$, $\mathbf{b} = (b_1, b_2, b_3)$, and $\mathbf{c} = (c_1, c_2, c_3)$, then

$$\mathbf{a}.(\mathbf{b}+\mathbf{c}) = \mathbf{a}.\mathbf{b} + \mathbf{a}.\mathbf{c}; \tag{2.17}$$

this is the *distributive law*. It is easily proved, for we have

$$
\begin{aligned}
\mathbf{a}.(\mathbf{b}+\mathbf{c}) &= (a_1, a_2, a_3).(b_1+c_1, b_2+c_2, b_3+c_3) \\
&= (a_1 b_1 + a_1 c_1 + a_2 b_2 + a_2 c_2 + a_3 b_3 + a_3 c_3) \\
&= (a_1 b_1 + a_2 b_2 + a_3 b_3) + (a_1 c_1 + a_2 c_2 + a_3 c_3) \\
&= \mathbf{a}.\mathbf{b} + \mathbf{a}.\mathbf{c}.
\end{aligned}
$$

Scalar invariants. Any scalar which takes the same value in each co-ordinate system with which it may be associated is called a *scalar invariant*. Thus, the components of a vector $\mathbf{a} = (a_1, a_2, a_3)$ are *not* scalar invariants because they may take different values in different coordinate systems. However the magnitude of **a**, i.e. $a = \sqrt{(a_1^2 + a_2^2 + a_3^2)}$, is a scalar invariant.

Since $\mathbf{a}.\mathbf{a} = a^2$, the scalar product of **a** with itself is a scalar invariant. This is a special case of the following more general result.

Scalar products are scalar invariants. To prove this, let (a_1', a_2', a_3'), (b_1', b_2', b_3') be the components of **a**, **b** relative to the axes $Ox'y'z'$. From equation (2.2),

$$a_i' = l_{ij} a_j \quad \text{and} \quad b_i' = l_{ik} b_k \quad (i = 1, 2, 3),$$

and so

$$a_i' b_i' = l_{ij} l_{ik} a_j b_k.$$

But, from the orthonormality conditions (1.29),

$$l_{ij} l_{ik} = \delta_{jk}.$$

Hence

$$
\begin{aligned}
a_i' b_i' &= \delta_{jk} a_j b_k \\
&= a_k b_k;
\end{aligned}
$$

i.e. $a_1' b_1' + a_2' b_2' + a_3' b_3' = a_1 b_1 + a_2 b_2 + a_3 b_3$. This shows that the scalar product of **a** and **b** is invariant under a rotation of the axes. Since the scalar product is also obviously invariant under a translation of the axes (for the components of **a**, **b** are themselves invariant under such a translation) it now follows that it is a scalar invariant.

Geometrical representation. Let two non-zero vectors $\mathbf{a} = (a_1, a_2, a_3)$, $\mathbf{b} = (b_1, b_2, b_3)$ be represented by $\overrightarrow{OA}$, $\overrightarrow{OB}$, and let the direction cosines relative to the axes $Oxyz$ of the directed line segments OA and OB be

α_1, α_2, α_3 and β_1, β_2, β_3 respectively. Then, since the coordinates of A, B are (a_1, a_2, a_3), (b_1, b_2, b_3), we have by equations (1.4)

$$\alpha_i = \frac{a_i}{a}, \quad \beta_i = \frac{b_i}{b} \quad (i = 1, 2, 3).$$

Thus, using the summation convention,

$$ab\alpha_i\beta_i = a_i b_i.$$

But if θ is the angle between OA, OB (Fig. 17), equation (1.9) shows that

$$\cos \theta = \alpha_i \beta_i.$$

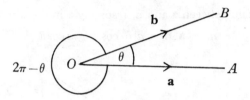

FIG. 17

Hence

$$\mathbf{a}.\mathbf{b} = ab \cos \theta. \tag{2.18}$$

This is an important relationship, often used as the definition of scalar product.

Note that since $\cos(2\pi - \theta) = \cos \theta$, no ambiguity arises if $2\pi - \theta$ is taken as the angle between OA, OB.

Two non-zero vectors $\mathbf{a}$ and $\mathbf{b}$ are at right angles to each other if and only if $\mathbf{a}.\mathbf{b} = 0$. For, if $\mathbf{a}$ and $\mathbf{b}$ are at right angles, $\theta = \frac{1}{2}\pi$ (or $\frac{3}{2}\pi$), and so by equation (2.18) $\mathbf{a}.\mathbf{b} = 0$. Also if $\mathbf{a}.\mathbf{b} = 0$ and $a \neq 0$, $b \neq 0$, then $\cos \theta = 0$; hence $\theta = \frac{1}{2}\pi$ (or $\frac{3}{2}\pi$), showing that $\mathbf{a}$ and $\mathbf{b}$ are at right angles.

Scalar products of pairs of $\mathbf{i}, \mathbf{j}, \mathbf{k}$. The unit vectors $\mathbf{i}, \mathbf{j}, \mathbf{k}$, introduced in the previous section, are such that the scalar product of any one with itself is unity and the scalar product of any one with any other is zero; for these vectors are each of unit magnitude and are mutually perpendicular. Thus,

$$\mathbf{i}.\mathbf{i} = \mathbf{j}.\mathbf{j} = \mathbf{k}.\mathbf{k} = 1;$$

and

$$\mathbf{i}.\mathbf{j} = \mathbf{j}.\mathbf{k} = \mathbf{k}.\mathbf{i} = 0. \tag{2.19}$$

Using these relations and the distributive law, the scalar product of two vectors

$$\mathbf{a} = a_1 \mathbf{i} + a_2 \mathbf{j} + a_3 \mathbf{k} \quad \text{and} \quad \mathbf{b} = b_1 \mathbf{i} + b_2 \mathbf{j} + b_3 \mathbf{k}$$

may be evaluated as follows:

$$\begin{aligned}
\mathbf{a}.\mathbf{b} &= (a_1\mathbf{i}+a_2\mathbf{j}+a_3\mathbf{k}).(b_1\mathbf{i}+b_2\mathbf{j}+b_3\mathbf{k}) \\
&= a_1b_1\mathbf{i}.\mathbf{i}+a_1b_2\mathbf{i}.\mathbf{j}+a_1b_3\mathbf{i}.\mathbf{k} \\
&\quad +a_2b_1\mathbf{j}.\mathbf{i}+a_2b_2\mathbf{j}.\mathbf{j}+a_2b_3\mathbf{j}.\mathbf{k} \\
&\quad +a_3b_1\mathbf{k}.\mathbf{i}+a_3b_2\mathbf{k}.\mathbf{j}+a_3b_3\mathbf{k}.\mathbf{k} \\
&= a_1b_1+a_2b_2+a_3b_3.
\end{aligned}$$

Of course, the definition (2.15) gives this at once.

Resolution of vectors. The *resolute* of a vector **a** in the direction of (or along) a unit vector $\hat{\mathbf{n}}$ is defined as

$$a_n = \mathbf{a}.\hat{\mathbf{n}}. \tag{2.20}$$

If θ is the angle between **a** and $\hat{\mathbf{n}}$, resolving **a** in the direction of $\hat{\mathbf{n}}$ gives

$$a_n = \mathbf{a}.\hat{\mathbf{n}} = a|\hat{\mathbf{n}}|\cos\theta = a\cos\theta.$$

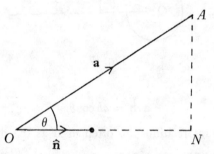

FIG. 18. The resolute of **a** in the direction $\hat{\mathbf{n}}$ is the sensed projection ON of **a** on $\hat{\mathbf{n}}$

Fig. 18 shows the geometrical interpretation of this. The resolute of **a** in the direction of $\hat{\mathbf{n}}$ is the sensed projection ON of **a** on $\hat{\mathbf{n}}$ (produced if necessary).

The resolute of **a** along any non-zero vector **b** is defined as $\mathbf{a}.\hat{\mathbf{b}}$, where

$$\hat{\mathbf{b}} = \mathbf{b}/b = (b_1, b_2, b_3)/(b_1^2+b_2^2+b_3^2)^{\frac{1}{2}}$$

is the unit vector in the same direction as **b**.

If

$$\mathbf{a} = a_1\mathbf{i}+a_2\mathbf{j}+a_3\mathbf{k},$$

then the resolutes of **a** in the directions **i**, **j**, **k** are respectively

$$\mathbf{a}.\mathbf{i} = a_1, \quad \mathbf{a}.\mathbf{j} = a_2, \quad \mathbf{a}.\mathbf{k} = a_3.$$

Thus the *resolutes* of **a** along the x-, y-, z-axes are identical to the corresponding components of **a**. It should be noted however that this result is not true if **a** is referred to a non-orthogonal coordinate system (see Exercise 45, § 2.8).

EXERCISES

21. If **a** and **b** are vectors as given in the table below, verify that the scalar products are as shown. Which of the three pairs of vectors is perpendicular?

	a	b	a.b
(i)	$(1,-1, \ 0)$	$(3, \ 4, \ 5)$	-1
(ii)	$(4, \ 1,-3)$	$(-1, \ 3,-7)$	20
(iii)	$(3, \ 1, \ 4)$	$(2,-2,-1)$	$0.$

22. Using the formula $\mathbf{a}.\mathbf{b}=ab\cos\theta$, find the angle between the vectors $\mathbf{a}=(0,-1,1)$, $\mathbf{b}=(3,4,5)$.

23. A set of rectangular cartesian axes is so arranged that the x-axis points East, the y-axis points North, and the z-axis points vertically upwards. Evaluate the scalar products of the vectors **a** and **b** in the following cases:
   (i) **a** is of magnitude 3 and points S.E., **b** is of magnitude 2 and points E.
   (ii) **a** is of unit magnitude and points N.E., **b** is of magnitude 2 and points vertically upwards;
   (iii) **a** is of unit magnitude and points N.E., **b** is of magnitude 2 and points W.

24. If
$$\mathbf{a} = \mathbf{i}-\mathbf{j}, \quad \mathbf{b} = -\mathbf{j}+2\mathbf{k},$$
show that
$$(\mathbf{a}+\mathbf{b}).(\mathbf{a}-2\mathbf{b}) = -9.$$

25. Show that the vectors $\mathbf{i}+\mathbf{j}+\mathbf{k}$, $\lambda^2\mathbf{i}-2\lambda\mathbf{j}+\mathbf{k}$ are perpendicular if and only if $\lambda=1$.

26. Find the component of **i** in the direction of the vector $\mathbf{i}+\mathbf{j}+2\mathbf{k}$.

27. Resolve the vector $3\mathbf{i}+4\mathbf{j}$ in the directions of the vectors $4\mathbf{i}-3\mathbf{j}$, $4\mathbf{i}+3\mathbf{j}$, and **k**.

28. If **a** and **b** are such that $a=b$, use the diagram below to interpret geometrically the relation
$$(\mathbf{a}+\mathbf{b}).(\mathbf{a}-\mathbf{b}) = a^2-b^2 = 0.$$

29. Prove vectorially that the perpendiculars onto the sides of a triangle from the opposite vertices are concurrent. [*Hint.* Draw the perpendiculars from the vertices A, B of a triangle ABC, and let them intersect at O. Let the position vectors of A, B, C relative to O be $\mathbf{a}$, $\mathbf{b}$, $\mathbf{c}$. Show that

$$\mathbf{a}.(\mathbf{b}-\mathbf{c}) = \mathbf{b}.(\mathbf{c}-\mathbf{a}) = 0.$$

Deduce that $\mathbf{c}.(\mathbf{a}-\mathbf{b})=0$, and give the interpretation.]

2.7   Vector products

The *vector product* of a vector $\mathbf{a}=(a_1,a_2,a_3)$ with a vector $\mathbf{b}=(b_1,b_2,b_3)$ is defined as

$$\mathbf{a} \times \mathbf{b} = (a_2 b_3 - a_3 b_2, a_3 b_1 - a_1 b_3, a_1 b_2 - a_2 b_1). \tag{2.21}$$

Alternatively, the vector product of

$$\mathbf{a} = a_1\mathbf{i}+a_2\mathbf{j}+a_3\mathbf{k} \quad \text{and} \quad \mathbf{b}=b_1\mathbf{i}+b_2\mathbf{j}+b_3\mathbf{k}$$

is defined to be

$$\mathbf{a} \times \mathbf{b} = \begin{vmatrix} \mathbf{i} & \mathbf{j} & \mathbf{k} \\ a_1 & a_2 & a_3 \\ b_1 & b_2 & b_3 \end{vmatrix}. \tag{2.22}$$

The notation used in (2.21) anticipates that the vector product of two vectors is a vector, and this result will now be established.

It should be noted first that $\mathbf{a} \times \mathbf{b}$ clearly satisfies conditions (i) and (ii) of the definition of a vector, given in § 2.2, and so it only remains to be shown that (iii) is satisfied.

For convenience, put

$$\begin{aligned} \mathbf{a} \times \mathbf{b} &= (a_2 b_3 - a_3 b_2, a_3 b_1 - a_1 b_3, a_1 b_2 - a_2 b_1) \\ &= (c_1, c_2, c_3). \end{aligned}$$

Referred to the axes $Ox'y'z'$ (defined in § 2.2),

$$\mathbf{a} = (a'_1, a'_2, a'_3), \quad \mathbf{b} = (b'_1, b'_2, b'_3),$$

and so, relative to these axes,

$$\begin{aligned} \mathbf{a} \times \mathbf{b} &= (a'_2 b'_3 - a'_3 b'_2, a'_3 b'_1 - a'_1 b'_3, a'_1 b'_2 - a'_2 b'_1) \\ &= (c'_1, c'_2, c'_3), \text{ say.} \end{aligned}$$

Condition (iii) will be satisfied if it can be proved that

$$\begin{aligned} c'_1 &= l_{11}c_1 + l_{12}c_2 + l_{13}c_3, \\ c'_2 &= l_{21}c_1 + l_{22}c_2 + l_{23}c_3, \\ c'_3 &= l_{31}c_1 + l_{32}c_2 + l_{33}c_3. \end{aligned} \tag{2.23}$$

Using the transformation law (2.1) for the vectors **a**, **b**, the quantity $c_1' = a_2' b_3' - a_3' b_2'$ becomes

$$
\begin{aligned}
c_1' &= (l_{21}a_1 + l_{22}a_2 + l_{23}a_3)(l_{31}b_1 + l_{32}b_2 + l_{33}b_3) \\
&\quad - (l_{31}a_1 + l_{32}a_2 + l_{33}a_3)(l_{21}b_1 + l_{22}b_2 + l_{23}b_3) \\
&= (l_{22}l_{33} - l_{23}l_{32})(a_2b_3 - a_3b_2) \\
&\quad + (l_{23}l_{31} - l_{21}l_{33})(a_3b_1 - a_1b_3) \\
&\quad + (l_{21}l_{32} - l_{22}l_{31})(a_1b_2 - a_2b_1).
\end{aligned}
$$

Substituting the results of Exercise 24, p. 13, and using the definitions of c_1, c_2, c_3, this reduces to

$$
c_1' = l_{11}c_1 + l_{12}c_2 + l_{13}c_3,
$$

which is the first of relations (2.23). The other two relations may be verified in a similar way, and the proof that $\mathbf{a} \times \mathbf{b}$ is a vector is then complete.

Interchanging **a** and **b** in (2.21) gives

$$
\mathbf{b} \times \mathbf{a} = (b_2 a_3 - b_3 a_2,\ b_3 a_1 - b_1 a_3,\ b_1 a_2 - b_2 a_1);
$$

thus

$$
\mathbf{b} \times \mathbf{a} = -\mathbf{a} \times \mathbf{b}, \tag{2.24}
$$

showing that *the operation of vector multiplication is non-commutative. It is essential therefore to preserve the order of the vectors in a vector product.*

If **a**, **b**, **c** are any three vectors, it is easily verified that

$$
\mathbf{a} \times (\mathbf{b} + \mathbf{c}) = \mathbf{a} \times \mathbf{b} + \mathbf{a} \times \mathbf{c}; \tag{2.25}
$$

thus the vector product obeys the *distributive law.*

Vector products of pairs of **i**, **j**, **k**. Since

$$
\mathbf{i} = (1, 0, 0), \quad \mathbf{j} = (0, 1, 0), \quad \mathbf{k} = (0, 0, 1),
$$

we have

$$
\mathbf{i} \times \mathbf{i} = \begin{vmatrix} \mathbf{i} & \mathbf{j} & \mathbf{k} \\ 1 & 0 & 0 \\ 1 & 0 & 0 \end{vmatrix} = \mathbf{0},
$$

$$
\mathbf{i} \times \mathbf{j} = \begin{vmatrix} \mathbf{i} & \mathbf{j} & \mathbf{k} \\ 1 & 0 & 0 \\ 0 & 1 & 0 \end{vmatrix} = \mathbf{k},
$$

and four similar relations obtained by cyclic permutation of **i**, **j**, **k**. In all

$$
\left. \begin{aligned}
\mathbf{i} \times \mathbf{i} &= \mathbf{j} \times \mathbf{j} = \mathbf{k} \times \mathbf{k} = \mathbf{0}, \\
\mathbf{i} \times \mathbf{j} &= \mathbf{k}, \quad \mathbf{j} \times \mathbf{k} = \mathbf{i}, \quad \mathbf{k} \times \mathbf{i} = \mathbf{j}.
\end{aligned} \right\} \tag{2.26}
$$

These identities should be compared with the corresponding identities (2.19) involving *scalar* products of pairs of the vectors **i**, **j**, **k**. Note that because of (2.24), interchanging the two vectors on the left-hand side of any one of the second set of identities (2.26) changes the sign in that identity; for example, $\mathbf{j} \times \mathbf{i} = -\mathbf{k}$.

Equations (2.26) may be used in conjunction with the distributive law (2.25) to evaluate vector products. For example, if

$$\mathbf{u} = \mathbf{i} + 3\mathbf{j} + \mathbf{k}, \quad \mathbf{v} = 2\mathbf{i} - \mathbf{j} + 2\mathbf{k},$$

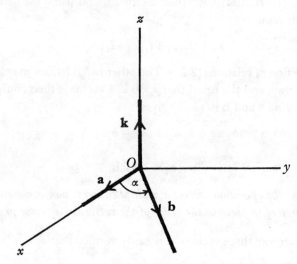

FIG. 19. The vector product $\mathbf{a} \times \mathbf{b}$ is of magnitude $ab \sin \alpha$ and in the direction of **k**

then

$$
\begin{aligned}
\mathbf{u} \times \mathbf{v} &= (\mathbf{i} + 3\mathbf{j} + \mathbf{k}) \times (2\mathbf{i} - \mathbf{j} + 2\mathbf{k}) \\
&= (2\mathbf{i} \times \mathbf{i} - \mathbf{i} \times \mathbf{j} + 2\mathbf{i} \times \mathbf{k}) + (6\mathbf{j} \times \mathbf{i} - 3\mathbf{j} \times \mathbf{j} + 6\mathbf{j} \times \mathbf{k}) \\
&\quad + (2\mathbf{k} \times \mathbf{i} - \mathbf{k} \times \mathbf{j} + 2\mathbf{k} \times \mathbf{k}) \\
&= (-\mathbf{k} - 2\mathbf{j}) + (-6\mathbf{k} + 6\mathbf{i}) + (2\mathbf{j} + \mathbf{i}) \\
&= 7\mathbf{i} - 7\mathbf{k}.
\end{aligned}
$$

It is however much more convenient to use (2.22). Thus

$$\mathbf{u} \times \mathbf{v} = \begin{vmatrix} \mathbf{i} & \mathbf{j} & \mathbf{k} \\ 1 & 3 & 1 \\ 2 & -1 & 2 \end{vmatrix} = 7\mathbf{i} - 7\mathbf{k}.$$

Geometrical interpretation of the vector product. Let **a**, **b** be given vectors, and choose rectangular axes $Oxyz$ such that **a**, **b** are parallel to the xy-plane and Ox is in the same direction as **a**. Let α be the angle between **a** and **b**, measured in the sense turning from Ox into the positive quadrant of the xy-plane (Fig. 19). Then

$$\mathbf{a} = a_1 \mathbf{i}, \quad \mathbf{b} = b_1 \mathbf{i} + b_2 \mathbf{j}$$

giving

$$\mathbf{a} \times \mathbf{b} = a_1 b_2 \mathbf{k}.$$

Also, for this special configuration,

$$a_1 = a, \quad b_2 = b \sin \alpha.$$

Hence

$$\mathbf{a} \times \mathbf{b} = ab \sin \alpha \, \mathbf{k}. \qquad (2.27)$$

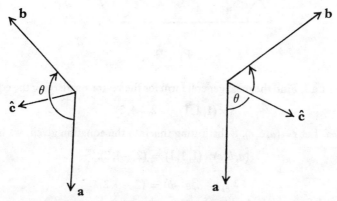

FIG. 20. The angle θ is always chosen in the range $0 \leqslant \theta \leqslant \pi$. In each diagram, $\hat{\mathbf{c}}$ is perpendicular to both $\mathbf{a}$ and $\mathbf{b}$

If $0 < \alpha < \pi$, $\sin \alpha > 0$, and so $\mathbf{a} \times \mathbf{b}$ is in the direction Oz; and if $\pi < \alpha < 2\pi$, $\sin \alpha < 0$, so that $\mathbf{a} \times \mathbf{b}$ is in the opposite direction to Oz. However, if the angle between $\mathbf{a}$ and $\mathbf{b}$ is defined suitably, the geometrical interpretation may be expressed more conveniently, as follows.

Let θ be the angle between $\mathbf{a}$ and $\mathbf{b}$, measured in the sense turning from $\mathbf{a}$ to $\mathbf{b}$ *and chosen so that* $0 \leqslant \theta \leqslant \pi$. Then the vector product $\mathbf{a} \times \mathbf{b}$ is the vector $ab \sin \theta \, \hat{\mathbf{c}}$, where $\hat{\mathbf{c}}$ is a unit vector at right angles to both $\mathbf{a}$ and $\mathbf{b}$ and such that to an observer looking in the direction of $\hat{\mathbf{c}}$, the sense in which θ is measured is clockwise. Fig. 20 makes the situation clear. Since $0 \leqslant \theta \leqslant \pi$, the magnitude of $\mathbf{a} \times \mathbf{b}$ is $ab \sin \theta$, and its direction is that of $\hat{\mathbf{c}}$.

Two non-zero vectors $\mathbf{a}$, $\mathbf{b}$ *are parallel or anti-parallel if and only if* $\mathbf{a} \times \mathbf{b} = \mathbf{0}$. For, if $\mathbf{a}$ and $\mathbf{b}$ are parallel, $\theta = 0$; and if $\mathbf{a}$ and $\mathbf{b}$ are anti-parallel, $\theta = \pi$. In either case $\mathbf{a} \times \mathbf{b} = \mathbf{0}$, since $\sin \theta = 0$. Also, if $\mathbf{a} \times \mathbf{b} = \mathbf{0}$ and if $a \neq 0$, $b \neq 0$, then $\sin \theta = 0$; hence $\theta = 0$ or π, showing that $\mathbf{a}$ and $\mathbf{b}$ are either parallel or anti-parallel.

4

EXAMPLE 2. Show that the area of a parallelogram with adjacent sides **a** and **b** is $|\mathbf{a} \times \mathbf{b}|$.

Solution. Denote by θ the smaller angle between **a** and **b** (Fig. 21). Drop a perpendicular from the end of **b** onto **a**. This perpendicular will be of length $b \sin \theta$. Thus, the area of the parallelogram is base $\times$ height $= ab \sin \theta$, which is $|\mathbf{a} \times \mathbf{b}|$.

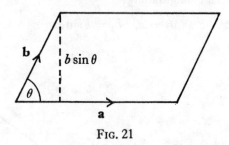

FIG. 21

EXAMPLE 3. Find the most general form for the vector **r** satisfying the equation

$$\mathbf{r} \times (1, 1, 1) = (2, -4, 2).$$

Solution. Let $\mathbf{r} = (a, b, c)$. Substituting this into the equation given, we have

$$(a, b, c) \times (1, 1, 1) = (2, -4, 2).$$

Thus

$$(b - c, c - a, a - b) = (2, -4, 2).$$

Using the definition of equality of vectors, this gives

$$b - c = 2,$$
$$c - a = -4,$$
$$a - b = 2.$$

These equations are not independent, but they are consistent; for adding the first two gives

$$b - a = -2,$$

which is the third equation.

Let

$$a = \lambda.$$

Then it follows at once that

$$b = \lambda - 2, \quad c = \lambda - 4.$$

Hence the general solution of the equation given may be represented in the form

$$\mathbf{r} = (\lambda, \lambda - 2, \lambda - 4),$$

where λ is arbitrary.

Remarks. If

$$\mathbf{r} \times \mathbf{a} = \mathbf{b},$$

the geometrical interpretation of vector products shows that **r** and **a** must both be perpendicular to **b**. Hence, if **a** and **b** are given, the equation can have no solution for **r** unless **a** is perpendicular to **b**.

It can be shown (see Exercise 54 at the end of this chapter) that the general solution for **r** is

$$\mathbf{r} = \lambda\mathbf{a} + (\mathbf{a} \times \mathbf{b})/a^2.$$

The reader may verify these remarks by reference to the example solved above.

EXAMPLE 4. Show vectorially that the bisectors of the angles of a triangle are concurrent.

Solution. Let the bisectors of the angles A and B of a triangle ABC intersect at O. Let the position vectors of A, B, C relative to O be **a**, **b**, **c**, respectively. Then

$$\overrightarrow{AC} = \mathbf{c} - \mathbf{a}, \quad \overrightarrow{CB} = \mathbf{b} - \mathbf{c}, \quad \overrightarrow{BA} = \mathbf{a} - \mathbf{b}.$$

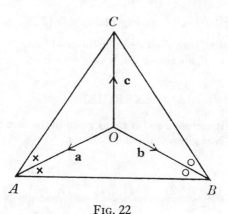

FIG. 22

Now if **û** and **v̂** are unit vectors, the internal bisector of the angle between them is parallel to the vector **û** + **v̂** (see Exercise 17). Unit vectors in the directions of $\overrightarrow{CA}$ and $\overrightarrow{BA}$ are

$$\frac{\mathbf{a} - \mathbf{c}}{|\mathbf{a} - \mathbf{c}|} \quad \text{and} \quad \frac{\mathbf{a} - \mathbf{b}}{|\mathbf{a} - \mathbf{b}|}.$$

Thus **a** is parallel to

$$\frac{\mathbf{a} - \mathbf{c}}{|\mathbf{a} - \mathbf{c}|} + \frac{\mathbf{a} - \mathbf{b}}{|\mathbf{a} - \mathbf{b}|}.$$

Similarly **b** is parallel to

$$\frac{\mathbf{b} - \mathbf{a}}{|\mathbf{b} - \mathbf{a}|} + \frac{\mathbf{b} - \mathbf{c}}{|\mathbf{b} - \mathbf{c}|}.$$

These conditions may be expressed as

$$\mathbf{a} \times \left[\frac{\mathbf{a} - \mathbf{c}}{|\mathbf{a} - \mathbf{c}|} + \frac{\mathbf{a} - \mathbf{b}}{|\mathbf{a} - \mathbf{b}|} \right] = 0$$

and

$$\mathbf{b} \times \left[\frac{\mathbf{b}-\mathbf{a}}{|\mathbf{b}-\mathbf{a}|} + \frac{\mathbf{b}-\mathbf{c}}{|\mathbf{b}-\mathbf{c}|} \right] = 0.$$

Since $\mathbf{a} \times \mathbf{a} = 0$ and $\mathbf{b} \times \mathbf{b} = 0$, these simplify to

$$\mathbf{a} \times \left[\frac{\mathbf{c}}{|\mathbf{a}-\mathbf{c}|} + \frac{\mathbf{b}}{|\mathbf{a}-\mathbf{b}|} \right] = 0 \qquad (2.28)$$

and

$$\mathbf{b} \times \left[\frac{\mathbf{a}}{|\mathbf{b}-\mathbf{a}|} + \frac{\mathbf{c}}{|\mathbf{b}-\mathbf{c}|} \right] = 0. \qquad (2.29)$$

The condition that CO should be the bisector of the angle at C can now be written down (by cyclic symmetry) as

$$\mathbf{c} \times \left[\frac{\mathbf{b}}{|\mathbf{c}-\mathbf{b}|} + \frac{\mathbf{a}}{|\mathbf{c}-\mathbf{a}|} \right] = 0. \qquad (2.30)$$

The result (2.30) follows by adding (2.28) and (2.29), and observing that:

(i) $|\mathbf{a}-\mathbf{c}| = |\mathbf{c}-\mathbf{a}|$, with similar results for $|\mathbf{a}-\mathbf{b}|$ and $|\mathbf{b}-\mathbf{c}|$; and
(ii) $\mathbf{a} \times \mathbf{b} + \mathbf{b} \times \mathbf{a} = 0$, $\mathbf{a} \times \mathbf{c} = -\mathbf{c} \times \mathbf{a}$, $\mathbf{b} \times \mathbf{c} = -\mathbf{c} \times \mathbf{b}$.

EXERCISES

30. Show that, if vectors $\mathbf{a}$ and $\mathbf{b}$ are as given in the first two columns of the table below, $\mathbf{a} \times \mathbf{b}$ is as given in the third column.

	$\mathbf{a}$	$\mathbf{b}$	$\mathbf{a} \times \mathbf{b}$
(i)	$(3, \ 7, \ 2)$	$(1, 3, 1)$	$(1, \ -1, \ 2)$
(ii)	$(1, -3, \ 0)$	$(-2, 5, 0)$	$(0, \ \ 0, -1)$
(iii)	$(8, \ 8, -1)$	$(5, 5, 2)$	$(21, -21, \ 0)$.

31. Axes $Oxyz$ are positioned so that Ox points East, Oy points North, and Oz points vertically upwards. Find the vector product $\mathbf{a} \times \mathbf{b}$ of vectors $\mathbf{a}$ and $\mathbf{b}$ in the following cases:

(i) $\mathbf{a}$ is of unit magnitude and points E, $\mathbf{b}$ is of magnitude 2 and points 30° N of E;
(ii) $\mathbf{a}$ is of unit magnitude and points E, $\mathbf{b}$ is of magnitude 2 and points SW;
(iii) $\mathbf{a}$ is of unit magnitude and points vertically upwards, $\mathbf{b}$ is of unit magnitude and points NE.

32. By taking components, prove the distributive law for vector products, namely

$$\mathbf{a} \times (\mathbf{b}+\mathbf{c}) = \mathbf{a} \times \mathbf{b} + \mathbf{a} \times \mathbf{c}.$$

33. Prove that for any scalar λ,

$$\mathbf{a} \times (\lambda \mathbf{b}) = (\lambda \mathbf{a}) \times \mathbf{b} = \lambda(\mathbf{a} \times \mathbf{b}).$$

34. If
$$\mathbf{a} \times \mathbf{b} = \mathbf{a} - \mathbf{b},$$
prove that
$$\mathbf{a} = \mathbf{b}.$$

35. Find the most general form for the vector $\mathbf{u}$ satisfying the equation
$$\mathbf{u} \times (2, 1, -1) = (1, 0, 0) \times (2, 1, -1).$$

36. Find a, b if
$$(a\mathbf{i} + b\mathbf{j} + \mathbf{k}) \times (2\mathbf{i} + 2\mathbf{j} + 3\mathbf{k}) = \mathbf{i} - \mathbf{j}.$$

37. By constructing an example, show that in general the associative law for vector products does not hold; that is, there exist $\mathbf{a}$, $\mathbf{b}$, $\mathbf{c}$ such that
$$\mathbf{a} \times (\mathbf{b} \times \mathbf{c}) \neq (\mathbf{a} \times \mathbf{b}) \times \mathbf{c}.$$

38. Prove vectorially that the medians of a triangle are concurrent. [*Hint.* Let E, F, G be the mid-points of the sides BC, CA, AB of a triangle ABC. Let AE, BF meet at O, and denote the position vectors of A, B, C relative to O by $\mathbf{a}$, $\mathbf{b}$, $\mathbf{c}$. Find the position vectors of E and F, and deduce that
$$\mathbf{a} \times (\mathbf{b} + \mathbf{c}) = 0 \quad \text{and} \quad \mathbf{b} \times (\mathbf{c} + \mathbf{a}) = 0.$$
Hence show that $\mathbf{c} \times (\mathbf{a} + \mathbf{b}) = 0$, and deduce the result required.]

2.8 The triple scalar product

The scalar $\mathbf{a} \cdot (\mathbf{b} \times \mathbf{c})$ is called a triple scalar product. If
$$\mathbf{a} = (a_1, a_2, a_3), \quad \mathbf{b} = (b_1, b_2, b_3), \quad \mathbf{c} = (c_1, c_2, c_3),$$
we have
$$\mathbf{a} \cdot (\mathbf{b} \times \mathbf{c}) = \mathbf{a} \cdot \begin{vmatrix} \mathbf{i} & \mathbf{j} & \mathbf{k} \\ b_1 & b_2 & b_3 \\ c_1 & c_2 & c_3 \end{vmatrix}.$$
Thus
$$\mathbf{a} \cdot (\mathbf{b} \times \mathbf{c}) = \begin{vmatrix} a_1 & a_2 & a_3 \\ b_1 & b_2 & b_3 \\ c_1 & c_2 & c_3 \end{vmatrix}. \tag{2.31}$$

It is easily verified that
$$\mathbf{a} \cdot (\mathbf{b} \times \mathbf{c}) = (\mathbf{a} \times \mathbf{b}) \cdot \mathbf{c}; \tag{2.32}$$
that is, *the 'dot' and 'cross' may be interchanged in a triple scalar product.*

Geometrical interpretation. Consider the parallelepiped with adjacent edges representing vectors $\mathbf{a}$, $\mathbf{b}$, $\mathbf{c}$ as shown in Fig. 23, with $\mathbf{b}$ and $\mathbf{c}$ horizontal. The volume V of the parallelepiped is 'area of base × height'. That is
$$V = |(bc \sin \theta)(a \cos \phi)|,$$

where θ is the angle between $\mathbf{b}$ and $\mathbf{c}$ and ϕ is the angle between $\mathbf{a}$ and the upward vertical. But

$$\mathbf{b} \times \mathbf{c} = bc \sin \theta \, \mathbf{k},$$

where $\mathbf{k}$ is a unit vector vertically upwards. Also

$$\mathbf{a} . \mathbf{k} = a \cos \phi.$$

It follows that

$$V = |\mathbf{a}.(\mathbf{b} \times \mathbf{c})|. \tag{2.33}$$

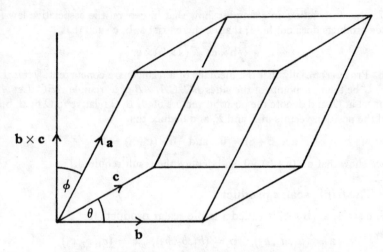

FIG. 23. The volume of a parallelepiped is $\mathbf{a}.(\mathbf{b} \times \mathbf{c})$

Condition for coplanar vectors. Three non-zero vectors $\mathbf{a}$, $\mathbf{b}$, $\mathbf{c}$ are coplanar if and only if

$$\mathbf{a}.(\mathbf{b} \times \mathbf{c}) = 0.$$

Proof. As the three vectors have non-vanishing magnitudes, the volume V of a parallelepiped with adjacent edges $\mathbf{a}$, $\mathbf{b}$, $\mathbf{c}$ is zero if and only if the vectors are coplanar. Thus (2.33) shows that $\mathbf{a}$, $\mathbf{b}$, $\mathbf{c}$ are coplanar if and only if $\mathbf{a}.(\mathbf{b} \times \mathbf{c}) = 0$.

Note. The triple scalar product $\mathbf{a}.(\mathbf{b} \times \mathbf{c})$ vanishes if any two of the three vectors are parallel or anti-parallel; but the converse of this statement is not true.

EXAMPLE 5. If the non-zero vectors $\mathbf{a}$, $\mathbf{b}$, $\mathbf{c}$ are not coplanar, show that any other vector $\mathbf{A}$ may be expressed uniquely in the form

$$\mathbf{A} = \lambda \mathbf{a} + \mu \mathbf{b} + \nu \mathbf{c},$$

where λ, μ, ν are scalars.

Solution. Let $\mathbf{A}=(A_1, A_2, A_3)$, $\mathbf{a}=(a_1, a_2, a_3)$, $\mathbf{b}=(b_1, b_2, b_3)$ and $\mathbf{c}=(c_1, c_2, c_3)$. Then

$$\mathbf{A} = \lambda\mathbf{a} + \mu\mathbf{b} + \nu\mathbf{c}$$

if and only if

$$A_1 = \lambda a_1 + \mu b_1 + \nu c_1,$$
$$A_2 = \lambda a_2 + \mu b_2 + \nu c_2,$$
$$A_3 = \lambda a_3 + \mu b_3 + \nu c_3.$$

These simultaneous equations for λ, μ, ν have a unique solution if and only if

$$\begin{vmatrix} a_1 & a_2 & a_3 \\ b_1 & b_2 & b_3 \\ c_1 & c_2 & c_3 \end{vmatrix} \neq 0,$$

i.e.

$$\mathbf{a}.(\mathbf{b} \times \mathbf{c}) \neq 0.$$

This condition is satisfied because $\mathbf{a}$, $\mathbf{b}$, $\mathbf{c}$ are not coplanar and are non-zero, and so the required result follows.

EXERCISES

39. By writing out components, or otherwise, prove that

$$\mathbf{a}.(\mathbf{b} \times \mathbf{c}) = (\mathbf{a} \times \mathbf{b}).\mathbf{c}.$$

40. If (x, y, z) is any point on the plane through the points (x_1, y_1, z_1), (x_2, y_2, z_2) and the origin, show that

$$\begin{vmatrix} x & y & z \\ x_1 & y_1 & z_1 \\ x_2 & y_2 & z_2 \end{vmatrix} = 0.$$

41. Show that for all scalars λ,

$$(\mathbf{a} + \lambda\mathbf{b}).(\mathbf{b} \times \mathbf{c}) = \mathbf{a}.(\mathbf{b} \times \mathbf{c}).$$

42. Show that

$$(\mathbf{a} + \mathbf{b} + \mathbf{c}).(\mathbf{b} \times \mathbf{c}) = \mathbf{a}.(\mathbf{b} \times \mathbf{c}).$$

43. If

$$A_1 = \lambda_1\mathbf{a} + \mu_1\mathbf{b} + \nu_1\mathbf{c},$$
$$A_2 = \lambda_2\mathbf{a} + \mu_2\mathbf{b} + \nu_2\mathbf{c},$$
$$A_3 = \lambda_3\mathbf{a} + \mu_3\mathbf{b} + \nu_3\mathbf{c},$$

show that

$$\mathbf{A}_1.(\mathbf{A}_2 \times \mathbf{A}_3) = \begin{vmatrix} \lambda_1 & \mu_1 & \nu_1 \\ \lambda_2 & \mu_2 & \nu_2 \\ \lambda_3 & \mu_3 & \nu_3 \end{vmatrix} \mathbf{a}.(\mathbf{b} \times \mathbf{c}).$$

44. Show that, given any four non-zero vectors $\mathbf{a}$, $\mathbf{b}$, $\mathbf{c}$, $\mathbf{d}$, there exist scalars p, q, r, s, not all zero, such that

$$p\mathbf{a} + q\mathbf{b} + r\mathbf{c} + s\mathbf{d} = \mathbf{0}.$$

[*Hint*. Consider the cases (i) when three of the vectors are not coplanar, (ii) when all four vectors are coplanar.]

45. Let OX, OY, OZ be a system of oblique rectilinear axes (i.e. axes such that OX, OY, OZ are straight lines which are not mutually perpendicular and not coplanar), and let $\mathbf{I}, \mathbf{J}, \mathbf{K}$ denote unit vectors in the three coordinate directions. If a vector $\mathbf{A}$ is expressed in the form

$$\mathbf{A} = A_1\mathbf{I} + A_2\mathbf{J} + A_3\mathbf{K},$$

then A_1, A_2, A_3 are called the *components* of $\mathbf{A}$. Show that these components are not identical with the *resolutes* of $\mathbf{A}$ along OX, OY, OZ.

2.9 The triple vector product

Vectors such as $(\mathbf{a} \times \mathbf{b}) \times \mathbf{c}$ and $\mathbf{a} \times (\mathbf{b} \times \mathbf{c})$ are called triple vector products. The following identities (proved below) are often needed:

$$(\mathbf{a} \times \mathbf{b}) \times \mathbf{c} = (\mathbf{a}.\mathbf{c})\mathbf{b} - (\mathbf{b}.\mathbf{c})\mathbf{a}; \tag{2.34}$$

$$\mathbf{a} \times (\mathbf{b} \times \mathbf{c}) = (\mathbf{a}.\mathbf{c})\mathbf{b} - (\mathbf{a}.\mathbf{b})\mathbf{c}. \tag{2.35}$$

Proof. Choose axes $Oxyz$ with the x-axis in the same direction as $\mathbf{a}$, and such that $\mathbf{b}$ is parallel to the xy-plane (as in Fig. 19, § 2.7). Then

$$\mathbf{a} = (a_1, 0, 0), \quad \mathbf{b} = (b_1, b_2, 0), \quad \mathbf{c} = (c_1, c_2, c_3).$$

Thus,

$$\mathbf{a} \times \mathbf{b} = (0, 0, a_1 b_2),$$

giving

$$(\mathbf{a} \times \mathbf{b}) \times \mathbf{c} = (-a_1 b_2 c_2, a_1 b_2 c_1, 0). \tag{2.36}$$

Also

$$(\mathbf{a}.\mathbf{c})\mathbf{b} - (\mathbf{b}.\mathbf{c})\mathbf{a} = a_1 c_1 \mathbf{b} - (b_1 c_1 + b_2 c_2)\mathbf{a}$$
$$= (-a_1 b_2 c_2, a_1 b_2 c_1, 0). \tag{2.37}$$

Comparing (2.36) and (2.37), the identity (2.34) is established.

The second identity (2.35) is easily proved likewise, or by making use of (2.34).

Remark. To remember (2.34) and (2.35), observe that each of the vectors *inside* the brackets on the left appears once *outside* the brackets on the right; and the 'middle' vector $\mathbf{b}$ appears first. Each term contains $\mathbf{a}, \mathbf{b}, \mathbf{c}$ once only.

EXERCISES

46. Adjacent sides of a triangle represent vectors $\mathbf{a}$ and $\mathbf{b}$. Show that the area of the triangle is $\frac{1}{2}|\mathbf{a} \times \mathbf{b}|$.

47. Establish formula (2.35) by making use of (2.34). [*Hint*. Use the result that for any two vectors $\mathbf{A}$ and $\mathbf{B}$, $\mathbf{A} \times \mathbf{B} = -\mathbf{B} \times \mathbf{A}$.]

48. Prove that, if $\mathbf{a}$, $\mathbf{b}$, $\mathbf{c}$ are non-zero and

$$(\mathbf{a} \times \mathbf{b}) \times \mathbf{c} = \mathbf{a} \times (\mathbf{b} \times \mathbf{c}),$$

then either (i) $\mathbf{b}$ is perpendicular to both $\mathbf{a}$ and $\mathbf{c}$, or (ii) $\mathbf{a}$ and $\mathbf{c}$ are parallel or anti-parallel. [*Hint.* Expand using (2.34) and (2.35).]

2.10 Products of four vectors

It is sometimes necessary to manipulate products of four vectors. This often involves the formulae (2.34) and (2.35), together with the knowledge that the dot and cross in the triple scalar product are always interchangeable. Thus, for example

$$(\mathbf{a} \times \mathbf{b}) . (\mathbf{c} \times \mathbf{d}) = \mathbf{a} . [\mathbf{b} \times (\mathbf{c} \times \mathbf{d})]. \tag{2.38}$$

Expanding the triple vector product gives

$$\mathbf{b} \times (\mathbf{c} \times \mathbf{d}) = (\mathbf{b} . \mathbf{d}) \mathbf{c} - (\mathbf{b} . \mathbf{c}) \mathbf{d}.$$

Substituting into equation (2.38) we obtain

$$(\mathbf{a} \times \mathbf{b}) . (\mathbf{c} \times \mathbf{d}) = (\mathbf{b} . \mathbf{d})(\mathbf{a} . \mathbf{c}) - (\mathbf{b} . \mathbf{c})(\mathbf{a} . \mathbf{d}),$$

so that

$$(\mathbf{a} \times \mathbf{b}) . (\mathbf{c} \times \mathbf{d}) = \begin{vmatrix} \mathbf{a} . \mathbf{c} & \mathbf{a} . \mathbf{d} \\ \mathbf{b} . \mathbf{c} & \mathbf{b} . \mathbf{d} \end{vmatrix}. \tag{2.39}$$

Other exercises on products of four vectors are given below.

EXERCISES

49. Show that

$$|\mathbf{a} \times \mathbf{b}|^2 = \mathbf{a}^2 \mathbf{b}^2 - (\mathbf{a} . \mathbf{b})^2.$$

50. Given two vectors $\mathbf{a}$ and $\mathbf{r}$ through the origin, show on a diagram the vector $(\mathbf{a} \times \mathbf{r}) \times \mathbf{a}$. Deduce that the length of the perpendicular onto $\mathbf{a}$ from the point with position vector $\mathbf{r}$ is

$$|\mathbf{a} \times \mathbf{r}|^2 / |(\mathbf{a} \times \mathbf{r}) \times \mathbf{a}|.$$

51. If $\hat{\mathbf{a}}$ and $\hat{\mathbf{b}}$ are unit vectors, show that

$$|\hat{\mathbf{a}} \times \hat{\mathbf{b}}|^2 = 1 - (\hat{\mathbf{a}} . \hat{\mathbf{b}})^2.$$

Show that this is another form of the trigonometrical identity $\sin^2\theta = 1 - \cos^2\theta$.

52. Show that

$$\mathbf{a} \times [\mathbf{b} \times (\mathbf{c} \times \mathbf{a})] = (\mathbf{a} . \mathbf{b}) \mathbf{a} \times \mathbf{c}.$$

2.11 Bound vectors

In mechanics, the *point of application* of a force or its *line of action*, may be important; the force, together with its point of application or its line of action, is then sometimes said to be a *bound vector*. We shall not enter into a detailed discussion, which belongs more properly to a course on mechanics.

EXERCISES

53. Let $\mathscr{L}$ be the line of action of a force $\mathbf{F}$, and let O be a given point. Then the *moment* of $\mathbf{F}$ about O is defined as

$$\mathbf{G} = \mathbf{r} \times \mathbf{F},$$

where $\mathbf{r} = \overrightarrow{OP}$ is the position vector of any point P on the line $\mathscr{L}$. Show that $\mathbf{G}$ is independent of the particular point P chosen on $\mathscr{L}$.

54. As in Exercise 53 above, the moment $\mathbf{G}$ of a force $\mathbf{F}$ about a point O is given by the equation

$$\mathbf{G} = \mathbf{r} \times \mathbf{F},$$

where $\mathbf{r}$ is the position vector relative to O of any point on $\mathscr{L}$ the line of action of $\mathbf{F}$. By direct substitution (or otherwise) show that the position vectors of points on $\mathscr{L}$ are given by

$$\mathbf{r} = \lambda \mathbf{F} + (\mathbf{F} \times \mathbf{G})/F^2,$$

where λ is a parameter. What is the perpendicular distance of O from $\mathscr{L}$?

CHAPTER 3

VECTOR FUNCTIONS OF A REAL VARIABLE.
DIFFERENTIAL GEOMETRY OF CURVES

3.1 Vector functions and their geometrical representation

The reader should already be familiar with the idea of a real function $f(x)$, say, of a real variable x. In this chapter the properties of *vector* functions $\mathbf{F}(t)$ of a real variable t will be discussed.

Suppose that the components of the vector

$$\mathbf{F}(t) = (f_1(t), f_2(t), f_3(t)) \tag{3.1}$$

are single-valued functions of a real variable t. Then $\mathbf{F}(t)$ is called a *vector function* of t. In most applications, t is a continuous variable and $f_1(t), f_2(t), f_3(t)$ are continuous[1] over some interval of t. If this is so, $\mathbf{F}(t)$ is said to be a *continuous vector function* of t. Examples of such functions are

and
$$\mathbf{F}(t) = (2, t^{\frac{1}{2}}, \sin t) \qquad 0 \leqslant t < \infty,$$

$$\mathbf{F}(t) = \begin{cases} (t^3, t, 3) & \text{for} \quad -\infty < t \leqslant 2, \\ (2t^2, 2, 6t^{-1}) & \text{for} \quad 2 < t < \infty. \end{cases}$$

The vector function

$$\mathbf{F}(t) = (1, t, t^{-1}) \qquad -1 \leqslant t \leqslant 1$$

is *not* continuous, because as t increases through zero the z-component t^{-1} changes value from $-\infty$ to ∞.

Geometrical representation of vector functions. Let a continuous vector function $\mathbf{F}(t)$ be represented by the position vector $\overrightarrow{OP}$, where O is the origin and P is the point $(f_1(t), f_2(t), f_3(t))$. Then, as t varies over its permissible range of values, P describes a *continuous curve* (in three dimensions, see Fig. 24). It is clear that in general both the magnitude and direction of

---

[1] Roughly speaking, a function is continuous if its value does not change suddenly at any point. However, for a precise definition of continuity in an interval see, for example, G. H. Hardy: *Pure Mathematics* (Cambridge, 1952), p. 186.

$\mathbf{F}(t)$ will vary with t. (A vector is constant only if both its magnitude and direction are fixed.) The equation

$$\overrightarrow{OP} = \mathbf{r} = \mathbf{F}(t), \tag{3.2}$$

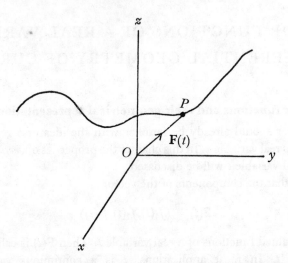

FIG. 24. A curve in three dimensions described by a point P whose position is given by an equation of the type (3.2)

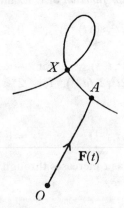

FIG. 25. A curve which intersects itself at X

where $\mathbf{r} = (x, y, z)$, is called the *parametric equation* of the curve described by P.

It should be noted that, although $\mathbf{F}(t)$ is taken to be a single-valued function of t, two (or possibly more) values of t may correspond to the same vector $\mathbf{F}$: in other words, there may fail to be a one–one correspondence between the vectors $\mathbf{F}(t)$ and the variable t. A simple instance of such a situation occurs when $\mathbf{F} = \overrightarrow{OA}$ is the position vector of a point A in motion; in this case t may be taken to denote time. If the point describes a curve which intersects itself at a point X, as in Fig. 25, then there will be two times t_1, t_2, say, at which A coincides with X. Thus

$$\mathbf{F}(t_1) = \mathbf{F}(t_2) = \overrightarrow{OX}.$$

A similar situation arises when a point retraces part (or all) of its path.

EXAMPLE 1. A point P has position vector

$$\overrightarrow{OP} = a(\cos\theta, 0, \sin\theta)$$

relative to rectangular cartesian axes $Oxyz$. Find the locus of P as θ varies and a remains constant.

Solution. Taking components,

$$x = a\cos\theta, \quad y = 0, \quad z = a\sin\theta.$$

Observing that $\cos^2\theta + \sin^2\theta = 1$, it is seen that the required locus is

$$x^2 + z^2 = a^2, \quad y = 0.$$

Thus, as θ varies, P traces out the circle $x^2 + z^2 = a^2$ in the zx-plane.

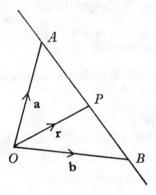

FIG. 26

EXAMPLE 2. Let $\mathbf{a}$, $\mathbf{b}$ be the position vectors relative to the origin O of points A, B. Show that the equation of the straight line through A, B can be expressed in the form

$$\mathbf{r} = \mathbf{a} + (\mathbf{b} - \mathbf{a})t, \tag{3.3}$$

where t is a parameter (Fig. 26).

Solution. The position vector of the point B relative to A is

$$\overrightarrow{AB} = \mathbf{b} - \mathbf{a}.$$

The point P with position vector $\mathbf{r}$ lies on the line through A and B (Fig. 26) if and only if

$$\overrightarrow{AP} = (\mathbf{b}-\mathbf{a})\,t,$$

where t is some real number. Noting that

$$\overrightarrow{OP} = \overrightarrow{OA}+\overrightarrow{AP},$$

we have

$$\mathbf{r} = \mathbf{a}+(\mathbf{b}-\mathbf{a})\,t.$$

This is the parametric equation of the straight line through A and B, because the position vectors of all points P on the line can be represented in this form.

Remark. If A is the point (x_0, y_0, z_0), B is the point (x_1, y_1, z_1), and P is the point (x, y, z), then the components of equation (3.3) are

$$x = x_0+(x_1-x_0)\,t, \quad y = y_0+(y_1-y_0)\,t, \quad z = z_0+(z_1-z_0)\,t.$$

Eliminating t gives

$$\frac{x-x_0}{x_1-x_0} = \frac{y-y_0}{y_1-y_0} = \frac{z-z_0}{z_1-z_0}.$$

This is the common form of the equation in rectangular cartesian coordinates of the straight line through $A(x_0, y_0, z_0)$ and $B(x_1, y_1, z_1)$.

EXERCISES

1. The continuous parameter t can take all real values. Sketch the curves whose parametric equations are respectively:

(i) $\mathbf{r} = (2\cos \pi t, \sin \pi t, 0)$,

(ii) $\mathbf{r} = (\sin \pi t, 0, 0)$,

(iii) $\mathbf{r} = (t, |t|, 0)$,

(iv) $\mathbf{r} = (t^2, t^3-t, 0)$,

(v) $\mathbf{r} = \begin{cases} (t, -t, 0) & \text{for } -\infty < t \leqslant 0 \\ (t, -t^2, 0) & \text{for } 0 \leqslant t < \infty. \end{cases}$

2. The points P and Q have position vectors

$$\mathbf{r}_P = (s^2+c, s, 1),$$
$$\mathbf{r}_Q = (2t, t, t),$$

where s and t are parameters and c is a constant. Find the value of c for which the loci of P and Q intersect, and show that the point of intersection is then $(2, 1, 1)$. What are the geometric forms of the two loci?

3. Show that the straight lines whose parametric equations are

$$\mathbf{r} = (1, 2, 5)+\lambda(0, 1, 0),$$

and

$$\mathbf{r} = (0, -2, 4)+\mu(1, 2, 1),$$

where λ, μ are parameters, have one point in common. Determine the coordinates of this point.

4. If λ, μ are parameters, show that the curves with parametric equations

$$\mathbf{r} = (1+\lambda, 1+2\lambda, 1+\lambda)$$

and

$$\mathbf{r} = (2\mu, \mu, 2-4\mu)$$

are straight lines, and that they intersect at right angles.

5. The vectors $\mathbf{a}$, $\hat{\mathbf{u}}$, $\hat{\mathbf{v}}$ are constant, and s, t are parameters which take all real values. Show that the locus of a point P with position vector

$$\mathbf{r} = \mathbf{a} + s\hat{\mathbf{u}} + t\hat{\mathbf{v}}$$

(relative to the origin) is a plane through the point with position vector $\mathbf{a}$ and parallel to the plane of the vectors $\hat{\mathbf{u}}$, $\hat{\mathbf{v}}$.

6. If θ, ϕ are parameters taking all real values, show that the locus of a point P with position vector

$$\mathbf{r} = (\cos\theta, \sin\theta\cos\phi, \sin\theta\sin\phi)$$

is a sphere with centre at the origin and of unit radius.

3.2 Differentiation of vectors

Suppose that, in some interval of t, the functions $f_1(t)$, $f_2(t)$, $f_3(t)$ are differentiable once with respect to t. Then the *first derivative of* $\mathbf{F}(t)$ is defined in the same interval to be

$$\frac{d\mathbf{F}}{dt} = \left(\frac{df_1}{dt}, \frac{df_2}{dt}, \frac{df_3}{dt}\right). \tag{3.4}$$

The derivative of a vector is also a vector. To establish this we must verify that conditions (i), (ii) and (iii) of the definition given in § 2.2 are satisfied.

It is at once obvious that the first condition is satisfied. Also, because $f_1(t)$, $f_2(t)$, $f_3(t)$ are unchanged by a translation of the axes, it follows that

$$df_1/dt, df_2/dt, df_3/dt$$

are unchanged by such a translation, and hence the second condition is satisfied.

To verify that condition (iii) of § 2.2 is satisfied, let the components of $\mathbf{F}$ relative to fixed axes $Ox_1'x_2'x_3'$ be (f_1', f_2', f_3'). As in Chapter 1 (§ 1.6), let l_{ij} denote the cosine of the angle between Ox_i' and Ox_j. Since the two sets of axes are taken to be fixed, each l_{ij} is independent of t, and hence differentiating the equations of transformation (2.2),

$$\frac{df_i'}{dt} = l_{ij}\frac{df_j}{dt}. \tag{3.5}$$

But df_1'/dt, df_2'/dt, df_3'/dt are the components of $d\mathbf{F}/dt$ referred to the new axes $Ox_1'x_2'x_3'$, and comparing (3.5) with (2.2) it is seen that condition (iii) holds. It follows that $d\mathbf{F}/dt$ is a vector.

The definition of higher-order derivatives of $\mathbf{F}$ presents no difficulty. For example, if f_1, f_2, f_3 are twice differentiable functions of t in some range, then the second derivative of $\mathbf{F}$ with respect to t is defined in that range as

$$\frac{d^2\mathbf{F}}{dt^2} \equiv \frac{d}{dt}\left(\frac{d\mathbf{F}}{dt}\right) \equiv \left(\frac{d^2f_1}{dt^2}, \frac{d^2f_2}{dt^2}, \frac{d^2f_3}{dt^2}\right). \tag{3.6}$$

By applying the theorem already proved to the vector $d\mathbf{F}/dt$, it follows at once that $d^2\mathbf{F}/dt^2$ is also a vector.

EXAMPLE 3. Find the values of λ for which the vector

$$\mathbf{A} = (\cos\lambda x, \sin\lambda x, 0)$$

satisfies the differential equation

$$\frac{d^2\mathbf{A}}{dx^2} = -9\mathbf{A}.$$

Solution. Using the formula for differentiation,

$$\frac{d\mathbf{A}}{dx} = (-\lambda\sin\lambda x, \lambda\cos\lambda x, 0)$$

and

$$\frac{d^2\mathbf{A}}{dx^2} = (-\lambda^2\cos\lambda x, -\lambda^2\sin\lambda x, 0).$$

Thus, the given differential equation is satisfied if $\lambda^2 = 9$, that is if $\lambda = \pm 3$.

EXERCISES

7. Write down the derivatives $d\mathbf{r}/dt$ and $d^2\mathbf{r}/dt^2$ for the following vectors:

(i) $\mathbf{r} = (2\cos\pi t, \sin\pi t, 0)$,

(ii) $\mathbf{r} = (t, t, e^t)$,

(iii) $\mathbf{r} = (|t|, t, 0)$ $(t \neq 0)$.

8. Given that

$$\frac{d\mathbf{r}}{dt} = \{-e^{-t}(\cos t + \sin t), e^{-t}(\cos t - \sin t), 0\},$$

and that, when $t = 0$, $\mathbf{r} = (1, 0, 0)$, determine $\mathbf{r}$. Sketch the locus of the point with position vector $\mathbf{r}$ for values of $t \geqslant 0$.

9. Given that the general solution of the differential equation

$$\frac{d^2x}{dt^2} + \omega^2 x = 0 \quad (\omega \text{ constant})$$

is

$$x = A\cos\omega t + B\sin\omega t,$$

where A, B are arbitrary constants, show that the general solution of the differential equation

$$\frac{d^2\mathbf{r}}{dt^2} + \omega^2\mathbf{r} = 0$$

is

$$\mathbf{r} = \mathbf{A}\cos\omega t + \mathbf{B}\sin\omega t,$$

where $\mathbf{A}$, $\mathbf{B}$ are arbitrary constant vectors.

If the motion of a point is such that its position vector $\mathbf{r}$ satisfies the differential equation above, show that the motion is confined to one plane.

3.3 Differentiation rules

The rules for differentiation of sums and products of vector functions are similar to the corresponding rules for differentiation of ordinary functions. If λ, $\mathbf{a}$, $\mathbf{b}$ are differentiable functions of t, the following identities hold:

$$\frac{d}{dt}(\mathbf{a}+\mathbf{b}) \equiv \frac{d\mathbf{a}}{dt} + \frac{d\mathbf{b}}{dt}; \tag{3.7}$$

$$\frac{d}{dt}(\lambda\mathbf{a}) \equiv \frac{d\lambda}{dt}\mathbf{a} + \lambda\frac{d\mathbf{a}}{dt}; \tag{3.8}$$

$$\frac{d}{dt}(\mathbf{a}\cdot\mathbf{b}) \equiv \frac{d\mathbf{a}}{dt}\cdot\mathbf{b} + \mathbf{a}\cdot\frac{d\mathbf{b}}{dt}; \tag{3.9}$$

$$\frac{d}{dt}(\mathbf{a}\times\mathbf{b}) \equiv \frac{d\mathbf{a}}{dt}\times\mathbf{b} + \mathbf{a}\times\frac{d\mathbf{b}}{dt}. \tag{3.10}$$

These identities are easily established by writing the vectors in component form.

For example,

$$\frac{d}{dt}(\mathbf{a}\cdot\mathbf{b}) \equiv \frac{d}{dt}(a_1 b_1 + a_2 b_2 + a_3 b_3)$$

$$\equiv \frac{da_1}{dt}b_1 + \frac{da_2}{dt}b_2 + \frac{da_3}{dt}b_3 + a_1\frac{db_1}{dt} + a_2\frac{db_2}{dt} + a_3\frac{db_3}{dt}$$

$$\equiv \frac{d\mathbf{a}}{dt}\cdot\mathbf{b} + \mathbf{a}\cdot\frac{d\mathbf{b}}{dt},$$

which proves (3.9).

Note that in (3.10) the order of $\mathbf{a}$ and $\mathbf{b}$ must be strictly observed, because of the non-commutative property of the vector products.

5

EXAMPLE 4. Show that the first derivative of a unit vector $\hat{\mathbf{a}}(t)$ is always perpendicular to $\hat{\mathbf{a}}(t)$, provided that the derivative is not zero.

Solution. We have

$$\hat{\mathbf{a}}.\hat{\mathbf{a}} = 1,$$

and hence

$$\frac{d\hat{\mathbf{a}}}{dt}.\hat{\mathbf{a}}+\hat{\mathbf{a}}.\frac{d\hat{\mathbf{a}}}{dt} = 0,$$

which gives

$$\hat{\mathbf{a}}.d\hat{\mathbf{a}}/dt = 0.$$

Since neither $\hat{\mathbf{a}}$ nor $d\hat{\mathbf{a}}/dt$ is zero, it follows from this that they must be perpendicular.

EXERCISES

10. Establish the identities (3.7), (3.8), and (3.10) in the text. Also *verify* that (3.9) holds for the case in which

$$\mathbf{a} = (1, t, t^2), \quad \mathbf{b} = (t^2, t, 1).$$

11. By writing $\mathbf{r}=r\hat{\mathbf{r}}$ show that, for any differentiable vector function $\mathbf{r}=\mathbf{r}(t)$,

$$\frac{dr}{dt} = \hat{\mathbf{r}}.\frac{d\mathbf{r}}{dt}.$$

12. Prove that

$$\frac{d}{dt}\{(\mathbf{a}\times\mathbf{b}).\mathbf{c}\} \equiv \left(\frac{d\mathbf{a}}{dt}\times\mathbf{b}\right).\mathbf{c}+\left(\mathbf{a}\times\frac{d\mathbf{b}}{dt}\right).\mathbf{c}+(\mathbf{a}\times\mathbf{b}).\frac{d\mathbf{c}}{dt}.$$

13. If

$$\mathbf{a}\times\frac{d\mathbf{b}}{dt} = \mathbf{b}\times\frac{d\mathbf{a}}{dt}$$

for all values of t, what can be deduced about $\mathbf{a}$ and $\mathbf{b}$?

3.4 The tangent to a curve. Smooth, piecewise smooth, and simple curves

In this section and the two following sections we discuss some important concepts relating to curves.

Suppose that a continuous curve $\mathscr{C}$ is the locus of the point P whose position vector (relative to the origin O of fixed axes $Oxyz$) is

$$\overrightarrow{OP} = \mathbf{r} = \mathbf{r}(t)$$
$$= (x(t), y(t), z(t)). \tag{3.11}$$

Let P' be a particular point on $\mathscr{C}$ at which $d\mathbf{r}/dt$ exists and is not zero. Then, at this point, $d\mathbf{r}/dt$ *lies along the tangent to the curve and is directed in the sense in which the curve is described by P as t increases.* To show this, let $t = t'$ at the point P'. Then

$$\overrightarrow{OP'} = \mathbf{r}(t').$$

At P',

$$\frac{d\mathbf{r}}{dt} = \lim_{t \to t'} \left(\frac{x(t) - x(t')}{t - t'}, \frac{y(t) - y(t')}{t - t'}, \frac{z(t) - z(t')}{t - t'} \right)$$

$$= \lim_{t \to t'} \frac{\mathbf{r}(t) - \mathbf{r}(t')}{t - t'}$$

$$= \lim_{t \to t'} \frac{\overrightarrow{P'P}}{t - t'}.$$

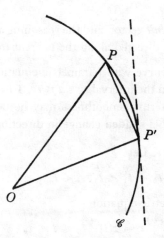

FIG. 27. As $t \to t'$, $\overrightarrow{P'P}/(t-t')$ ultimately lies along the tangent to $\mathscr{C}$ at P'

It is clear that as $t \to t'$, P approaches P' and so $\overrightarrow{P'P}/(t-t')$ ultimately lies along the tangent at P' (Fig. 27). Thus $d\mathbf{r}/dt$ is directed along the tangent to $\mathscr{C}$.

To determine the sense in which $d\mathbf{r}/dt$ points, choose axes with origin O at P', and with the x-axis parallel to $d\mathbf{r}/dt$. Then at O

$$\frac{d\mathbf{r}}{dt} = \left(\frac{dx}{dt}, 0, 0 \right).$$

As t increases, the point P moves through O in the positive or negative x-direction according as dx/dt is greater than or less than zero. It follows that $d\mathbf{r}/dt$ points in the direction in which P moves along $\mathscr{C}$ as t increases.

The unit tangent. Suppose that at the point P', with parameter t', on the curve with parametric equation

$$\mathbf{r} = \mathbf{r}(t) \qquad t_0 \leqslant t \leqslant t_1, \tag{3.12}$$

$d\mathbf{r}/dt$ exists and is not zero[1]. Then the vector

$$\hat{\mathbf{T}} = \frac{d\mathbf{r}/dt}{|d\mathbf{r}/dt|} \tag{3.13}$$

is defined as the *unit tangent* at P'. If $|d\mathbf{r}/dt| \to 0$ or ∞ as $t \to t'$, then we define

$$\hat{\mathbf{T}} = \lim_{t \to t'} \frac{d\mathbf{r}/dt}{|d\mathbf{r}/dt|}, \tag{3.14}$$

provided the limit exists.

It is clear that $\hat{\mathbf{T}}$ is a *unit* vector, and the reasoning at the beginning of this section shows that it is directed along the tangent to the curve.

Smooth curve. The curve with parametric equation (3.12) is said to be *smooth* if, at all points in the interval $t_0 \leqslant t \leqslant t_1$, $\hat{\mathbf{T}}$ exists and is continuous. In rather less precise terms, smoothness may be taken to mean that the curve does not undergo a sudden change in direction at any point.

Piecewise smooth curve. Let

$$t_0 < t_1 < t_2 \ldots < t_{n-1} < t_n.$$

The curve with parametric equation

$$\mathbf{r} = \mathbf{r}(t) \quad t_0 \leqslant t \leqslant t_n$$

is said to be *piecewise smooth* if (i) $\mathbf{r}(t)$ is continuous in the interval $t_0 \leqslant t \leqslant t_n$, and (ii) the unit tangent $\hat{\mathbf{T}}$ is continuous in the interval $t_0 \leqslant t \leqslant t_n$ except at the points $t_1, t_2, \ldots, t_{n-1}$. Thus a piecewise smooth curve consists of a finite number of smooth curves, linked end to end (Fig. 28(b)).

Simple open curve. A piecewise smooth curve

$$\mathbf{r} = \mathbf{r}(t) \quad t_0 \leqslant t \leqslant t_n$$

is said to be *simple and open* if each point on it corresponds to just one value of t. Thus a simple open curve does not cross or meet itself at any point.

Simple closed curve. A piecewise smooth curve with parametric equation

---

[1] For the definition of a derivative at an end point of an interval see, for example, G. H. Hardy: *Pure Mathematics* (Cambridge, 1952), p. 286.

$$\mathbf{r} = \mathbf{r}(t) \quad t_0 \leqslant t \leqslant t_n$$

is said to be *simple and closed* if its end points (corresponding to $t = t_0$ and $t = t_n$) coincide, and each other point corresponds to just one value of t.

An elementary example of a simple closed curve is the unit circle in the xy-plane, with parametric equation

$$\mathbf{r} = (\cos t, \sin t, 0) \quad 0 \leqslant t \leqslant 2\pi.$$

The reader should verify that this satisfies all the conditions of the above definition. Observe also that if the range of t were $0 \leqslant t \leqslant 4\pi$, the circle would be described twice as t covered the range, and so there would no longer be a simple curve.

Some examples of the various possible types of curve are shown in Fig. 28.

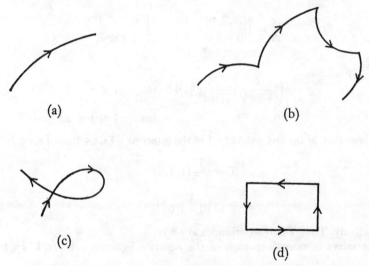

(a) (b) (c) (d)

FIG. 28. Classification of curves : (a) Smooth curve. Also a simple curve if covered once only. (b) Piecewise smooth. Simple if covered once only. (c) Smooth curve. Not simple. (d) Simple closed curve (if covered once only)

Change of parameter. The curve with parametric equation

$$\mathbf{r} = \mathbf{r}(t) \quad t_0 \leqslant t \leqslant t_1$$

has a *sense* (or *orientation*) defined as the direction in which it is described as t increases from t_0 to t_1. The sense is often indicated in a diagram by an arrow on the curve (Fig. 28). It is desirable that under a change of parameter the sense of description should be preserved. To ensure this, the parametric transformation

$$t = t(u)$$

is only considered *allowable* if it is such that dt/du is non-negative at all points in the interval $u_0 \leqslant u \leqslant u_1$, where u_0, u_1 are the values of u corresponding to t_0, t_1, respectively. With this restriction, u is nowhere decreasing as t is increasing, and hence the sense of description of the curve is preserved.

EXAMPLE 5. Show that the unit tangent to the curve

$$\mathbf{r} = \begin{cases} (t^2, 2t, 0) & -1 \leqslant t \leqslant 1 \\ (1, 4-2t, 0) & 1 \leqslant t \leqslant 2 \end{cases}$$

is discontinuous at the point $t=1$. Verify that the curve is piecewise smooth, and indicate its sense in a diagram.

Solution. We have

$$\frac{d\mathbf{r}}{dt} = \begin{cases} (2t, 2, 0) & \text{for} \quad -1 \leqslant t \leqslant 1 \\ (0, -2, 0) & \text{for} \quad 1 \leqslant t \leqslant 2. \end{cases}$$

Thus

$$\hat{\mathbf{T}} = \begin{cases} \left(\dfrac{t}{(1+t^2)^{\frac{1}{2}}}, \dfrac{1}{(1+t^2)^{\frac{1}{2}}}, 0 \right) & \text{for} \quad -1 \leqslant t \leqslant 1 \\ (0, -1, 0) & \text{for} \quad 1 \leqslant t \leqslant 2. \end{cases}$$

It follows that, at the end points $t=1$ of the intervals $-1 \leqslant t \leqslant 1$ and $1 \leqslant t \leqslant 2$,

$$\hat{\mathbf{T}} = \frac{1}{\sqrt{2}}(1, 1, 0)$$

and

$$\hat{\mathbf{T}} = (0, -1, 0),$$

respectively. Thus $\hat{\mathbf{T}}$ is discontinuous at $t=1$.

The curve is smooth in each of the separate intervals $-1 \leqslant t \leqslant 1$, $1 \leqslant t \leqslant 2$,

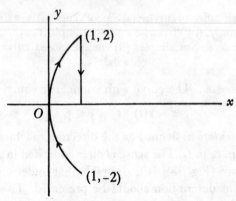

FIG. 29. A curve consisting of part of the parabola $y^2 = 4x$ and the straight line $x = 1$

because throughout each interval $\hat{\mathbf{T}}$ is continuous. Further, $\mathbf{r}(t)$ is clearly continuous at the point t equals 1. It follows that the curve is piecewise smooth.

The curve lies entirely in the xy-plane. In the interval $-1 \leqslant t \leqslant 1$ it consists of part of the parabola $y^2 = 4x$; and in the interval $1 \leqslant t \leqslant 2$ it is part of the straight line $x = 1$ (Fig. 29). The sense of the curve is indicated by the arrows in the diagram.

EXERCISES

14. Show in a diagram the direction of $d\mathbf{r}/dt$ at the points $t=0$, $t=1$, $t=-1$ for the curves whose parametric equations are:

$$\text{(i)} \qquad \mathbf{r} = (2\cos \tfrac{1}{2}\pi t, \sin \tfrac{1}{2}\pi t, 0) \quad (-2 \leqslant t \leqslant 2),$$
$$\text{(ii)} \qquad \mathbf{r} = (t^2, t^3 - t, 0) \qquad\qquad (-\infty < t < \infty).$$

Classify the curves according to the definitions in this section.

15. Show that the unit tangent to the curve

$$\mathbf{r} = (3, t, t^2)$$

is

$$\hat{\mathbf{T}} = (0, 1, 2t)/(1 + 4t^2)^{\frac{1}{2}}.$$

16. Sketch the curve with parametric equation

$$\mathbf{r} = (t, |\sin t|, 0) \quad 0 \leqslant t \leqslant 3\pi,$$

and show that it is piecewise smooth.

17. Sketch the curve with parametric equation

$$\mathbf{r} = \begin{cases} (\sin t, \cos t, 0) & \text{for} \quad 0 \leqslant t \leqslant \tfrac{1}{2}\pi \\ (1, \tfrac{1}{2}\pi - t, 0) & \text{for} \quad \tfrac{1}{2}\pi \leqslant t \leqslant \tfrac{3}{2}\pi, \end{cases}$$

and show that it is smooth. [*Hint.* Show that $\mathbf{r}$ and the unit tangent $\hat{\mathbf{T}}$ are continuous at the point $t = \tfrac{1}{2}\pi$.]

3.5 Arc Length

Let

$$\mathbf{r} = \mathbf{r}(t) = (x(t), y(t), z(t)) \quad t_0 \leqslant t \leqslant t_1 \tag{3.15}$$

be the parametric equation of a piecewise smooth curve $\mathscr{C}$. Define

$$\frac{ds}{dt} = \left|\frac{d\mathbf{r}}{dt}\right| = (\dot{x}^2 + \dot{y}^2 + \dot{z}^2)^{\frac{1}{2}}, \tag{3.16}$$

where $\dot{x}$, $\dot{y}$ and $\dot{z}$ denote dx/dt, dy/dt and dz/dt, respectively. Then

$$s(t) = \int_{t_0}^{t} (\dot{x}^2 + \dot{y}^2 + \dot{z}^2)^{\frac{1}{2}} \, dt \tag{3.17}$$

is defined as the *arc length* of $\mathscr{C}$ from the point with parameter t_0 to the (variable) point with parameter t. The total length of the curve is defined as $s(t_1) = l$. Observe that $s(t_0) = 0$.

At points where ds/dt is finite, non-zero and continuous, the element of arc ds corresponding to an increment dt in t is defined as

$$ds = (\dot{x}^2 + \dot{y}^2 + \dot{z}^2)^{\frac{1}{2}} \, dt = (dx^2 + dy^2 + dz^2)^{\frac{1}{2}};$$

this has an obvious intuitive geometrical interpretation, as indicated in Fig. 30, and the motivation for the definitions (3.16) and (3.17) also becomes clear.

By the definition (3.16), $ds/dt \geqslant 0$, and hence the substitution $t = t(s)$ is an allowable change of parameter. In terms of s,

$$\mathbf{r} = \mathbf{r}(s) = (x(s), y(s), z(s)) \quad 0 \leqslant s \leqslant l, \tag{3.18}$$

and this is called the *intrinsic equation* of the curve.

The unit tangent $\hat{\mathbf{T}}$ may be obtained from the intrinsic equation quite simply. For, upon substituting (3.16), the expressions (3.13) and (3.14) both reduce to

$$\hat{\mathbf{T}} = \frac{d\mathbf{r}}{ds} = \left(\frac{dx}{ds}, \frac{dy}{ds}, \frac{dz}{ds} \right). \tag{3.19}$$

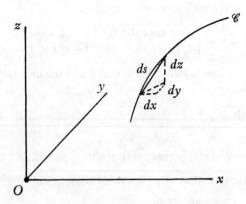

Fig. 30. Geometrical interpretation of the element of arc length
$$ds = \sqrt{(dx^2 + dy^2 + dz^2)}$$

EXERCISE

18. Find the intrinsic equation of the curve with parametric equation

$$\mathbf{r} = (a \cos t, a \sin t, bt) \quad 0 \leqslant t \leqslant 2\pi.$$

3.6   Curvature and torsion

Let $\hat{\mathbf{T}}$ be the unit tangent to the curve with intrinsic equation $\mathbf{r} = \mathbf{r}(s)$, and write

$$\frac{d\hat{\mathbf{T}}}{ds} = \kappa\hat{\mathbf{N}} \quad \kappa \geqslant 0, \tag{3.20}$$

where κ is a positive function of s and $\hat{\mathbf{N}}$ is a unit vector. The worked example at the end of § 3.3 shows that $\hat{\mathbf{N}}$ is perpendicular to $\hat{\mathbf{T}}$: it is called the *principal unit normal vector*. The proportionality factor κ is defined to be the *curvature*, and is a measure of the rate at which the direction of the tangent changes with s. For example, if the curve is a straight line, $\hat{\mathbf{T}}$ is constant in direction as well as in magnitude, and so $\kappa = 0$. The quantity $\rho = \kappa^{-1}$ is defined to be the *radius of curvature*.

Another vector of importance in the differential theory of curves is the *unit binormal vector*, defined as

$$\hat{\mathbf{B}} = \hat{\mathbf{T}} \times \hat{\mathbf{N}}. \tag{3.21}$$

The three unit vectors $\hat{\mathbf{T}}, \hat{\mathbf{N}}, \hat{\mathbf{B}}$ form an orthonormal right-handed triad (Fig. 31).

The derivative of $\hat{\mathbf{B}}$ with respect to s is parallel or antiparallel to $\hat{\mathbf{N}}$. To

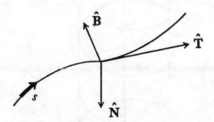

FIG. 31. The unit tangent $\hat{\mathbf{T}}$, unit principal normal $\hat{\mathbf{N}}$, and unit binormal $\hat{\mathbf{B}}$ at a particular point on a curve

show this, we first differentiate (3.21) and use (3.20), thus obtaining the equation

$$\frac{d\hat{\mathbf{B}}}{ds} = \kappa\hat{\mathbf{N}} \times \hat{\mathbf{N}} + \hat{\mathbf{T}} \times \frac{d\hat{\mathbf{N}}}{ds} = \hat{\mathbf{T}} \times \frac{d\hat{\mathbf{N}}}{ds}.$$

Now $d\hat{\mathbf{B}}/ds$ is normal to $\hat{\mathbf{B}}$ and hence lies in the plane of $\hat{\mathbf{N}}$ and $\hat{\mathbf{T}}$. But $\hat{\mathbf{T}} \times d\hat{\mathbf{N}}/ds$ is normal to $\hat{\mathbf{T}}$, and so

$$\frac{d\hat{\mathbf{B}}}{ds} = -\tau\hat{\mathbf{N}}, \tag{3.22}$$

where τ is a function of s. This proportionality factor is called the *torsion* of the curve, and is a measure of the rate at which the direction of the binormal changes with s.

For curves which lie in a plane Π, $\hat{\mathbf{T}}$ clearly lies in Π; therefore $\hat{\mathbf{N}}$ (which is proportional to the rate of change of $\hat{\mathbf{T}}$) also lies in Π. Thus, for a plane curve, $\hat{\mathbf{B}}$ is a constant vector at right angles to Π and the torsion is zero ($\tau = 0$).

EXAMPLE 6. Consider the *circular helix*, defined parametrically as

$$\mathbf{r} = (a\cos t, a\sin t, bt), \tag{3.23}$$

where a, b are constants. Equating the components,

$$x = a\cos t, \quad y = a\sin t, \quad z = bt,$$

and so for all t

$$x^2 + y^2 = a^2.$$

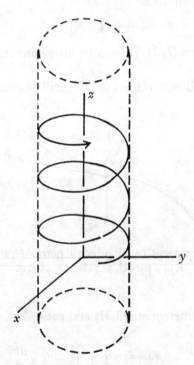

FIG. 32. Circular helix

The curve therefore lies on the surface of a circular cylinder of radius a and axis Oz. It spirals around the z-axis, as shown in Fig. 32. For this curve

$$\frac{ds}{dt} = \left\{ \left(\frac{dx}{dt}\right)^2 + \left(\frac{dy}{dt}\right)^2 + \left(\frac{dz}{dt}\right)^2 \right\}^{\frac{1}{2}} = (a^2 + b^2)^{\frac{1}{2}}.$$

Hence

$$\hat{\mathbf{T}} = \frac{d\mathbf{r}}{ds} = \frac{d\mathbf{r}}{dt}\frac{dt}{ds} = \frac{1}{(a^2+b^2)^{\frac{1}{2}}}(-a\sin t, a\cos t, b). \tag{3.24}$$

Also

$$\kappa\hat{\mathbf{N}} = \frac{d\hat{\mathbf{T}}}{ds} = \frac{-a}{a^2+b^2}(\cos t, \sin t, 0). \tag{3.25}$$

It follows that the principal unit normal $\hat{\mathbf{N}}$ is always parallel to the xy-plane. Also, equating the magnitudes of the two sides of (3.25), the curvature is

$$\kappa = |a|/(a^2+b^2). \tag{3.26}$$

From equations (3.24), (3.25),

$$\hat{\mathbf{B}} = \hat{\mathbf{T}}\times\hat{\mathbf{N}} = \frac{a}{\kappa(a^2+b^2)^{3/2}}(b\sin t, -b\cos t, a).$$

Thus

$$\tau\hat{\mathbf{N}} = -\frac{d\hat{\mathbf{B}}}{ds} = \frac{-ab}{\kappa(a^2+b^2)^2}(\cos t, \sin t, 0),$$

and using (3.25) this reduces to

$$\tau\hat{\mathbf{N}} = \frac{b}{a^2+b^2}\hat{\mathbf{N}}.$$

The torsion of the helix is therefore

$$\tau = b/(a^2+b^2). \tag{3.27}$$

Notice that when $b=0$ the curve reduces to a circle of radius $|a|$ in the xy-plane. As expected, the curvature is then $1/|a|$ (from (3.26)), and the torsion is zero (from (3.27)).

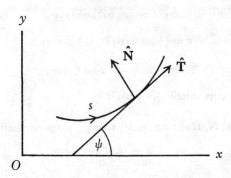

FIG. 33. The angle ψ is defined to be the angle between the unit tangent and the x-axis

EXAMPLE 7. As a second example we show that, for a plane curve in the xy-plane, if the unit tangent in the direction of s increasing makes an angle ψ with the positive x-axis, then the radius of curvature ρ is given by

$$\rho = |ds/d\psi|.$$

(This conclusion agrees with the usual elementary definition of ρ for a plane curve.)

Solution. In terms of the angle ψ (Fig. 33)

$$\hat{\mathbf{T}} = (\cos\psi, \sin\psi, 0),$$

and so

$$\frac{d\hat{\mathbf{T}}}{ds} = \frac{d\psi}{ds}\frac{d\hat{\mathbf{T}}}{d\psi}$$

$$= \frac{d\psi}{ds}(-\sin\psi, \cos\psi, 0).$$

It follows by comparison with the formula

$$\frac{d\hat{\mathbf{T}}}{ds} = \kappa\hat{\mathbf{N}}$$

that

$$\kappa = \left|\frac{d\psi}{ds}\right|.$$

Thus the radius of curvature is

$$\rho = \kappa^{-1} = \left|\frac{ds}{d\psi}\right|.$$

Note that

$$\hat{\mathbf{N}} = \begin{cases} (-\sin\psi, \cos\psi, 0) & \text{when } ds/d\psi > 0 \\ (\sin\psi, -\cos\psi, 0) & \text{when } ds/d\psi < 0. \end{cases}$$

EXERCISES

19. Show that the unit tangent vector to the curve

$$\mathbf{r} = (4\cos t, \cos 2t, 2t + \sin 2t)$$

is

$$\hat{\mathbf{T}} = (-\sin t, -\sin t \cos t, \cos^2 t).$$

Show also that the curvature is $\frac{1}{4}(1 + \cos^2 t)^{\frac{1}{2}}$.

20. By finding $\hat{\mathbf{T}}$, $\hat{\mathbf{N}}$, $\hat{\mathbf{B}}$ in turn, verify that the plane parabolic curve

$$\mathbf{r} = (t, \tfrac{1}{2}t^2, 0)$$

has zero torsion.

21. With the notation used in the text, show that

$$\frac{d\hat{\mathbf{N}}}{ds} = -\kappa\hat{\mathbf{T}} + \tau\hat{\mathbf{B}}.$$

[*Hint.* Differentiate the relation $\hat{\mathbf{N}} = \hat{\mathbf{B}} \times \hat{\mathbf{T}}$.]
Note. This result, together with the relations $d\hat{\mathbf{T}}/ds = \kappa\hat{\mathbf{N}}$ and $d\hat{\mathbf{B}}/ds = -\tau\hat{\mathbf{N}}$,

constitute the *Serret–Frenet formulae*. These formulae are fundamental to the differential geometry of curves.

3.7 Applications in kinematics

The components of acceleration of a point moving along a curve. When a particle (or a point P) is in motion, its position relative to a given coordinate system will depend upon time t. If $\mathbf{r} = \mathbf{r}(t)$ is the position vector of P, then the *velocity* $\mathbf{v}$ and the *acceleration* $\mathbf{f}$ relative to the chosen coordinate system are defined as

$$\mathbf{v} = \dot{\mathbf{r}}, \quad \mathbf{f} = \ddot{\mathbf{r}} = \dot{\mathbf{v}}, \tag{3.28}$$

where $\dot{\mathbf{r}}$ denotes $d\mathbf{r}/dt$, $\ddot{\mathbf{r}}$ denotes $d^2\mathbf{r}/dt^2$, etc.

Let $s = s(t)$ denote the arc length of the curve $\mathbf{r} = \mathbf{r}(t)$ that is covered by the particle at time t. Then

$$\mathbf{v} = \dot{\mathbf{r}} = \dot{s}\frac{d\mathbf{r}}{ds} = \dot{s}\hat{\mathbf{T}}, \tag{3.29}$$

where $\dot{s}$ is the *speed* of P and $\hat{\mathbf{T}}$ is the unit tangent vector directed, at every instant, in the sense in which P is moving. Also

$$\mathbf{f} = \frac{d}{dt}(\dot{s}\hat{\mathbf{T}}) = \ddot{s}\hat{\mathbf{T}} + \dot{s}^2\frac{d\hat{\mathbf{T}}}{ds}$$

$$= \ddot{s}\hat{\mathbf{T}} + \rho^{-1}\dot{s}^2\hat{\mathbf{N}}, \tag{3.30}$$

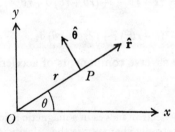

FIG. 34. The unit vectors $\hat{\mathbf{r}}$ and $\hat{\boldsymbol{\theta}}$ in planc polar coordinates r, θ

where $\hat{\mathbf{N}}$ is the principal unit normal of the curve described by P and ρ is the radius of curvature. The components of acceleration are therefore $\ddot{s}$ in the direction of the tangent vector and $\dot{s}^2/\rho$ along the *principal* normal. *This result is not restricted to plane motion.*

The components of acceleration in plane polar coordinates. Using the ideas introduced in this chapter, we can derive well-known formulae for components of velocity and acceleration in terms of plane polar coordinates (r, θ). Let $\hat{\mathbf{r}}, \hat{\boldsymbol{\theta}}$ denote unit vectors in the r, θ plane, such that $\hat{\mathbf{r}}$ points away from the origin O, and $\hat{\boldsymbol{\theta}}$ is normal to $\hat{\mathbf{r}}$ and in the direction θ increasing (Fig. 34).

We first prove two important subsidiary results, viz.

$$\frac{d\hat{\mathbf{r}}}{d\theta} = \hat{\boldsymbol{\theta}}, \quad \frac{d\hat{\boldsymbol{\theta}}}{d\theta} = -\hat{\mathbf{r}}. \tag{3.31}$$

Take rectangular cartesian axes: Ox in the direction $\theta = 0$, Oy in the direction $\theta = \frac{1}{2}\pi$, and Oz to complete a right-handed triad. Then

$$\hat{\mathbf{r}} = (\cos\theta, \sin\theta, 0), \quad \hat{\boldsymbol{\theta}} = (-\sin\theta, \cos\theta, 0).$$

Hence

$$\frac{d\hat{\mathbf{r}}}{d\theta} = (-\sin\theta, \cos\theta, 0), \quad \frac{d\hat{\boldsymbol{\theta}}}{d\theta} = -(\cos\theta, \sin\theta, 0),$$

and the required formulae (3.31) follow at once.

Now the position vector of P is

$$\mathbf{r} = r\hat{\mathbf{r}}. \tag{3.32}$$

Hence

$$\dot{\mathbf{r}} = \dot{r}\hat{\mathbf{r}} + r\dot{\theta}\frac{d\hat{\mathbf{r}}}{d\theta}$$

$$= \dot{r}\hat{\mathbf{r}} + r\dot{\theta}\hat{\boldsymbol{\theta}}. \tag{3.33}$$

The radial and transverse components of velocity are therefore $\dot{r}$, $r\dot{\theta}$ respectively. Also

$$\ddot{\mathbf{r}} = \ddot{r}\hat{\mathbf{r}} + \dot{r}\dot{\theta}\frac{d\hat{\mathbf{r}}}{d\theta} + (\dot{r}\dot{\theta} + r\ddot{\theta})\hat{\boldsymbol{\theta}} + r\dot{\theta}^2\frac{d\hat{\boldsymbol{\theta}}}{d\theta}$$

$$= (\ddot{r} - r\dot{\theta}^2)\hat{\mathbf{r}} + (2\dot{r}\dot{\theta} + r\ddot{\theta})\hat{\boldsymbol{\theta}}. \tag{3.34}$$

Thus the radial and transverse components of acceleration are $\ddot{r} - r\dot{\theta}^2$ and $2\dot{r}\dot{\theta} + r\ddot{\theta}$.

EXAMPLE 8. When an electron moves in a magnetic field it experiences a force $e\mathbf{v} \times \mathbf{B}$, where e is the electronic charge, $\mathbf{v}$ is the velocity, and $\mathbf{B}$ is the magnetic induction vector. Hence if $\mathbf{f}$ is the acceleration and m is the mass of the electron, its equation of motion is

$$m\mathbf{f} = e\mathbf{v} \times \mathbf{B}. \tag{3.35}$$

If $\mathbf{B}$ is uniform and independent of time t, show that the path described by the electron is a circular helix.

Solution. Let the z-axis be chosen in the direction of $\mathbf{B}$, so that $\mathbf{B} = B\mathbf{k}$, and let $\mathbf{r}$ be the position vector of the electron at time t. Then (3.35) becomes

$$\ddot{\mathbf{r}} = p\dot{\mathbf{r}} \times \mathbf{k}, \tag{3.36}$$

where $p = eB/m$ and dots denote differentiation with respect to t. Now

$$\mathbf{r} = x\mathbf{i} + y\mathbf{j} + z\mathbf{k},$$

and hence

$$\ddot{x}\mathbf{i}+\ddot{y}\mathbf{j}+\ddot{z}\mathbf{k} = p(\dot{x}\mathbf{i}+\dot{y}\mathbf{j}+\dot{z}\mathbf{k})\times\mathbf{k}.$$

Separating into components,

$$\ddot{x} = p\dot{y}, \quad \ddot{y} = -p\dot{x}, \quad \ddot{z} = 0. \tag{3.37}$$

Let the origin be chosen at the position of the particle when $t=0$, and choose the direction of the x-axis so that the initial velocity of the particle is $u\mathbf{i}+w\mathbf{k}$ (that is, so that the initial velocity component in the y-direction is zero). Then equations (3.37) have to be solved subject to the *initial conditions*

$$\dot{x} = u, \quad \dot{y} = 0, \quad \dot{z} = w, \tag{3.38}$$

and

$$x = y = z = 0. \tag{3.39}$$

The solution of the last of equations (3.37) is obtained at once as

$$z = wt. \tag{3.40}$$

Integrating the first two of equations (3.37) and using the initial conditions gives

$$\dot{x} = py+u, \quad \dot{y} = -px.$$

Substituting these results, the first two of equations (3.37) may now be rewritten as

$$\ddot{x}+p^2x = 0 \quad \text{and} \quad \ddot{y}+p^2y = -pu.$$

The general solutions for x, y are therefore

$$x = A\cos pt+B\sin pt,$$
$$y = C\cos pt+D\sin pt-u/p,$$

where A, B, C, D are arbitrary constants. Using the initial conditions (3.38), (3.39) it follows easily that $A=D=0$, $B=C=u/p$. Thus

$$x = \frac{u}{p}\sin pt, \quad y = \frac{u}{p}(\cos pt-1), \quad z = wt,$$

and these are the parametric equations of a circular helix with axis $x=0$, $y=-u/p$.

EXERCISES

22. At an origin O on the Earth's surface, the z-axis points vertically upwards. A particle moving under the influence of constant gravity only has acceleration

$$\frac{d^2\mathbf{r}}{dt^2} = (0,0,-g),$$

where $\mathbf{r}$ is the position vector and t denotes time. If the particle is projected from the origin when $t=0$ with velocity $(u,0,v)$, show by integrating the above differential equation that the locus of its path is

$$\mathbf{r} = (ut,0,vt-\tfrac{1}{2}gt^2).$$

23. A particle moves with velocity $\mathbf{v}$ and acceleration $\mathbf{f}$. Show that the radius of curvature of its path is

$$\rho = v^3/|\mathbf{v} \times \mathbf{f}|.$$

Use this formula to determine the radius of curvature at the origin of the path of a particle whose position vector at time t is

$$\mathbf{r} = (t, t^2, t^3).$$

24. A point moves so that its position vector $\mathbf{r}$ satisfies the differential equation

$$\frac{d^2\mathbf{r}}{dt^2} = \mathbf{g} - \lambda \frac{d\mathbf{r}}{dt},$$

where t denotes time, $\mathbf{g}$ is a constant vector, and λ is a constant scalar. If the point is at the origin at time $t=0$ and is then moving with velocity $\mathbf{u}$, show that

$$\mathbf{r} = \frac{\mathbf{g}}{\lambda^2}(\lambda t + e^{-\lambda t} - 1) + \frac{\mathbf{u}}{\lambda}(1 - e^{-\lambda t}).$$

[*Hint.* Show first that

$$\frac{d\mathbf{r}}{dt} + \lambda \mathbf{r} = \mathbf{g}t + \mathbf{u}.$$

Then multiply this equation by $e^{\lambda t}$ and integrate.]

CHAPTER 4

SCALAR AND VECTOR FIELDS

4.1 Regions

In the further development of vector analysis we shall be concerned with functions defined on certain point sets. It is convenient to introduce here some associated definitions.

Open region. A set of points constitutes an *open region* of three-dimensional space if: (i) each pair of points in the set can be joined by a continuous curve consisting entirely of points in the set, and (ii) each point is the centre of some sphere containing only points in the set.

Closed region. A set S of points constitutes a *closed region* if: (i) each pair of points in S can be joined by a continuous curve consisting entirely of points in S, and (ii) the points not in S form one or more open regions.

Boundary. A point P belonging to a closed region $\mathscr{R}$ is called a *boundary point* if every sphere centred at P contains at least one point which does not belong to $\mathscr{R}$. A set of boundary points of $\mathscr{R}$ forms a boundary or part of a boundary if each pair of points in the set can be joined by a continuous curve consisting entirely of boundary points.

The definitions above, applicable to point sets in three-dimensional space, can easily be modified to define open and closed regions and boundaries of sets of points in a plane: the sphere is simply replaced by its two-dimensional counterpart, the circle.

A simple example of an open region is the set of points (x, y, z) such that

$$x^2 + y^2 + z^2 < 1;$$

that is, the set of all points lying inside (but not on the surface of) the sphere of unit radius centred at the origin. However, the set of points (x, y, z) such that

$$x^2 + y^2 + z^2 \leqslant 1,$$

constitutes a closed region whose boundary is the sphere $x^2 + y^2 + z^2 = 1$.

EXERCISES

1. Does the set of all points in space constitute an open region or a closed region?

2. Consider the set of points (x, y) in the xy-plane such that

$$1 < x^2 + y^2 \leqslant 2.$$

Explain why this set constitutes neither an open region nor a closed region.

4.2 Functions of several variables

For the reader whose knowledge of functions of several independent variables is slight, we give a brief account in this section of most of the concepts and results needed later.[1] As we shall be concerned mainly with functions of three independent real variables, attention is confined to this case. However, most results quoted extend in an obvious way to functions of more than three variables.

Throughout this section, the independent variables x, y, z are rectangular cartesian coordinates in three-dimensional space.

Continuity. Let the real valued function $f(x, y, z)$ be defined at all points in some open region containing the point $P(a, b, c)$. Let $Q(x, y, z)$ be any other point in the region. We say that $f(x, y, z)$ is *continuous* at the point $P(a, b, c)$ if the difference

$$f(x, y, z) - f(a, b, c)$$

between the values of f at Q and P tends to zero whenever Q approaches P *along any path.*

Partial derivatives. The *first-order partial derivative* $\partial f / \partial x$ of the function $f(x, y, z)$ with respect to x is defined as

$$\lim_{h \to 0} \frac{f(x+h, y, z) - f(x, y, z)}{h}, \tag{4.1}$$

whenever the limit exists. It follows that $\partial f / \partial x$ is obtained by differentiating f with respect to x, treating y, z as constants. The partial derivatives of f with respect to y, z are defined similarly. There are thus three first-order partial derivatives of $f(x, y, z)$, viz.

$$\frac{\partial f}{\partial x}, \quad \frac{\partial f}{\partial y}, \quad \frac{\partial f}{\partial z}.$$

[1] Further information can be found in the following books: P. J. Hilton: *Partial Derivatives* (Routledge and Kegan Paul); and R. Courant: *Differential and Integral Calculus* (Blackie).

They are also sometimes denoted by

$$f_x, \quad f_y, \quad f_z.$$

EXAMPLE 1. Find f_x, f_y, f_z for the function

$$f = x^3 + x^2 y + xyz.$$

Solution. Treating y, z as constants and differentiating f with respect to x, we obtain immediately

$$f_x = 3x^2 + 2xy + yz.$$

Similarly, treating x, z as constants,

$$f_y = x^2 + xz,$$

and treating x, y as constants,

$$f_z = xy.$$

Suppose that the value of a partial derivative, say f_x, is required at a particular point (x_0, y_0, z_0). Then, apart from the obvious procedure of evaluating f_x and then putting $x = x_0$, $y = y_0$, $z = z_0$, we can also put $y = y_0$, $z = z_0$ in the function $f(x, y, z)$, then find the derivative with respect to x, and finally put $x = x_0$ in the result. Thus, in the example above, it is seen that

$$f_x(x_0, y_0, z_0) = 3x_0^2 + 2x_0 y_0 + y_0 z_0.$$

We also have

$$f(x, y_0, z_0) = x^3 + x^2 y_0 + xy_0 z_0.$$

Hence

$$f_x(x, y_0, z_0) = 3x^2 + 2xy_0 + y_0 z_0,$$

and so, as before,

$$f_x(x_0, y_0, z_0) = 3x_0^2 + 2x_0 y_0 + y_0 z_0.$$

Higher-order partial derivatives. It is clear that the first-order partial derivatives of $f(x, y, z)$ will themselves be functions of x, y, z. When these partial derivatives are differentiated partially with respect to x, y, or z, we obtain *second-order partial derivatives*. The partial derivatives of $\partial f / \partial x$ with respect to x, y, z are denoted respectively by

$$\frac{\partial^2 f}{\partial x^2}, \quad \frac{\partial^2 f}{\partial y \, \partial x}, \quad \frac{\partial^2 f}{\partial z \, \partial x},$$

or alternatively by

$$f_{xx}, \quad f_{yx}, \quad f_{zx}.$$

A similar notation is used for the second-order derivatives arising from $\partial f / \partial y$ and $\partial f / \partial z$.

Derivatives of higher order are defined in a similar manner, and an obvious extension of the above notation is used.

EXAMPLE 2. Find $\partial^3 f/\partial z^3$ for the function

$$f = (x+2y+3z)^4.$$

Solution. We find at once

$$\frac{\partial f}{\partial z} = 4(x+2y+3z)^3 \frac{\partial}{\partial z}(x+2y+3z)$$

$$= 12(x+2y+3z)^3.$$

Differentiating again with respect to z,

$$\frac{\partial^2 f}{\partial z^2} = 12 \times 3 \times 3(x+2y+3z)^2,$$

and hence

$$\frac{\partial^3 f}{\partial z^3} = 108 \times 2 \times 3(x+2y+3z)$$

$$= 648(x+2y+3z).$$

EXAMPLE 3. Find $\partial^2 f/\partial y\,\partial x$ and $\partial^2 f/\partial x\,\partial y$ for the function

$$f = \sin(ax+by+cz),$$

where a, b, c are constants.

Solution. Differentiating f partially with respect to x,

$$\frac{\partial f}{\partial x} = a\cos(ax+by+cz).$$

Hence, differentiating this with respect to y,

$$\frac{\partial^2 f}{\partial y\,\partial x} = -ab\sin(ax+by+cz).$$

Also, differentiating f partially with respect to y,

$$\frac{\partial f}{\partial y} = b\cos(ax+by+cz).$$

Hence,

$$\frac{\partial^2 f}{\partial x\,\partial y} = -ab\sin(ax+by+cz).$$

In the example above, it is observed that $f_{yx}=f_{xy}$, so the order in which the x-, y-differentiations are performed is in this case immaterial. The same situation arises for nearly all commonly occurring functions; in fact, if the reader were to write down a function $f(x,y,z)$ 'at random', it is most unlikely that pairs of mixed derivatives f_{xy}, f_{yx} would not be identical. For reference, we quote the following theorem which sets out sufficient conditions for the order of partial differentiation to be immaterial.

THEOREM. If all the mixed second-order derivatives (f_{xy}, f_{xz}, etc.) of the function $f(x,y,z)$ exist and are continuous at a given point, then, at that point,

$$f_{xy} = f_{yx}, \quad f_{yz} = f_{zy}, \quad f_{zx} = f_{xz}. \tag{4.2}$$

Continuously differentiable functions. The function $f(x,y,z)$ is said to be *continuously differentiable* in an open region $\mathcal{R}$ if its first-order partial derivatives f_x, f_y, f_z exist and are continuous at every point of $\mathcal{R}$.

Functions which are not continuously differentiable require special treatment and will not be discussed in this book. For emphasis, we shall occasionally remind the reader of this when stating important results, *but otherwise it will be assumed without comment that the functions discussed are properly defined and continuously differentiable in some open region. Furthermore, whenever we have occasion to introduce second- or higher-order derivatives of a function, we shall assume that they too exist and are continuous throughout any open region considered.*

The chain rule. Let F be a continuously differentiable function of f, g, h and suppose that each of f, g, h is a continuously differentiable function of x, y, z. Then, it may be shown that F is a continuously differentiable composite function of x, y, z and that

$$\frac{\partial F}{\partial x} = \frac{\partial F}{\partial f}\frac{\partial f}{\partial x} + \frac{\partial F}{\partial g}\frac{\partial g}{\partial x} + \frac{\partial F}{\partial h}\frac{\partial h}{\partial x}$$

$$\frac{\partial F}{\partial y} = \frac{\partial F}{\partial f}\frac{\partial f}{\partial y} + \frac{\partial F}{\partial g}\frac{\partial g}{\partial y} + \frac{\partial F}{\partial h}\frac{\partial h}{\partial y} \tag{4.3}$$

$$\frac{\partial F}{\partial z} = \frac{\partial F}{\partial f}\frac{\partial f}{\partial z} + \frac{\partial F}{\partial g}\frac{\partial g}{\partial z} + \frac{\partial F}{\partial h}\frac{\partial h}{\partial z}.$$

This is the rule for differentiation of a function of functions, usually called the *chain rule*.

To illustrate this result, let

$$F = f - 4g + h,$$

where

$$f = 3x^2 + 2y^2, \quad g = (x-z)^2, \quad h = y + 1.$$

Then,

$$\frac{\partial F}{\partial x} = \frac{\partial F}{\partial f}\frac{\partial f}{\partial x} + \frac{\partial F}{\partial g}\frac{\partial g}{\partial x} + \frac{\partial F}{\partial h}\frac{\partial h}{\partial x}$$

$$= 1 \times 3 \times 2x + (-4) \times 2(x-z) + 1 \times 0$$

$$= 8z - 2x.$$

Similarly $\partial F/\partial y$ and $\partial F/\partial z$ may also be evaluated using the chain rule, and without writing down F explicitly as a function of x, y, z.

EXERCISES

3. If
$$f(x,y) = ax^2 + 2hxy + by^2,$$
where a, b, h are constants, find f_x, f_y, f_{xx}, f_{yy}, f_{xy}, and f_{yx}.

4. If
$$f(x,y) = \sin xy,$$
verify that $f_{xy} = f_{yx}$.

5. If
$$r^2 = x^2 + y^2 + z^2,$$
find $\partial r/\partial x$, $\partial r/\partial y$ and $\partial r/\partial z$. Hence show that
$$\frac{\partial^2}{\partial x^2}\left(\frac{1}{r}\right) + \frac{\partial^2}{\partial y^2}\left(\frac{1}{r}\right) + \frac{\partial^2}{\partial z^2}\left(\frac{1}{r}\right) = 0,$$
except when $r=0$.

6. If
$$u = (Ar^n + Br^{-n})\cos n\theta,$$
where A, B and n are constants, verify that
$$\frac{\partial^2 u}{\partial r^2} + \frac{1}{r}\frac{\partial u}{\partial r} + \frac{1}{r^2}\frac{\partial^2 u}{\partial \theta^2} = 0.$$

7. Let u be a function of x and y which satisfies the differential equation
$$\frac{\partial u}{\partial x} + u = \frac{\partial^2 u}{\partial y^2}.$$
If $v = ue^x$, prove that
$$\frac{\partial v}{\partial x} = \frac{\partial^2 v}{\partial y^2}.$$

8. The function $u(x,t)$ satisfies the differential equation
$$\frac{\partial^2 u}{\partial x^2} = \frac{1}{c^2}\frac{\partial^2 u}{\partial t^2},$$
where c is a constant. If the independent variables are changed to
$$\xi = x + ct \quad \text{and} \quad \eta = x - ct,$$
prove that
$$\frac{\partial^2 u}{\partial \xi \, \partial \eta} = 0.$$

[*Hint.* Show first that
$$\partial u/\partial x = \partial u/\partial \xi + \partial u/\partial \eta$$

and

$$\partial u/\partial t = c\ \partial u/\partial \xi - c\ \partial u/\partial \eta.]$$

4.3　Definitions of scalar and vector fields

Suppose that a scalar $\Omega(x,y,z)$ is defined on a point set S in three-dimensional space; that is, to each point $P(x,y,z)$ in S there corresponds a value of $\Omega(x,y,z)$. Then Ω is called a *scalar function of position* or a *scalar field*. Likewise, if a vector $\mathbf{F}(x,y,z)$ is defined on the set S, then $\mathbf{F}$ is called a *vector function of position* or a *vector field*. In practice the set S will nearly always constitute a region, as defined in § 4.1, and accordingly we shall henceforth restrict our attention to this case. Since the point $P(x,y,z)$ is completely specified by its position vector $\mathbf{r} = (x,y,z)$ relative to the origin, the notation

$$\Omega = \Omega(\mathbf{r}), \quad \mathbf{F} = \mathbf{F}(\mathbf{r}) \tag{4.4}$$

may also be used to signify that Ω, $\mathbf{F}$ are functions of position.

Simple examples of scalar fields are

$$\Omega = x^2 + y^2 + z^2 \tag{4.5}$$

and

$$\Omega = 1/x. \tag{4.6}$$

Examples of vector fields are

$$\mathbf{F} = x\mathbf{i} + y\mathbf{j} + z\mathbf{k} \tag{4.7}$$

and

$$\mathbf{F} = (1 - x^2 - y^2 - z^2)^{\frac{1}{2}}(\mathbf{i} + \mathbf{j}). \tag{4.8}$$

The scalar field (4.5) and the vector field (4.7) are defined over the whole of space. The scalar field (4.6) is defined at all points except those lying on the plane $x = 0$; and the vector field (4.8) is defined only at points lying inside or on the sphere

$$x^2 + y^2 + z^2 = 1.$$

Scalar and vector fields arise naturally in a variety of physical situations. For example, when a gas flows along a pipe there are associated with any point in the pipe the gas pressure p, the density ρ, and the velocity $\mathbf{v}$ at that point: thus p and ρ are scalar fields, and $\mathbf{v}$ is a vector field, associated with the motion.

4.4　Gradient of a scalar field

If the scalar field $\Omega(x,y,z)$ is defined and continuously differentiable on some open region $\mathscr{R}$, then the *gradient* of Ω is defined as

$$\operatorname{grad}\Omega = \frac{\partial\Omega}{\partial x}\mathbf{i} + \frac{\partial\Omega}{\partial y}\mathbf{j} + \frac{\partial\Omega}{\partial z}\mathbf{k} = \left(\frac{\partial\Omega}{\partial x}, \frac{\partial\Omega}{\partial y}, \frac{\partial\Omega}{\partial z}\right). \tag{4.9}$$

The gradient of Ω is a vector field on $\mathscr{R}$.

Proof. At every point of $\mathscr{R}$, grad Ω clearly satisfies condition (i) of the definition of a vector, given in § 2.2. To complete the proof we must establish (a) that the components are invariant under a translation of the coordinate axes, and (b) that the components transform according to the vector law (see equations (2.1)) under a rotation of the axes.

(a) Consider a translation to new coordinate axes $O'XYZ$, such that the coordinates (X, Y, Z) are related to the original coordinates (x, y, z) by the equations

$$x = X + a, \quad y = Y + b, \quad z = Z + c, \tag{4.10}$$

where a, b, c are constants. By the chain rule (equations (4.3)),

$$\frac{\partial \Omega}{\partial X} = \frac{\partial x}{\partial X} \frac{\partial \Omega}{\partial x} + \frac{\partial y}{\partial X} \frac{\partial \Omega}{\partial y} + \frac{\partial z}{\partial X} \frac{\partial \Omega}{\partial z} = \frac{\partial \Omega}{\partial x}.$$

Similarly,

$$\frac{\partial \Omega}{\partial Y} = \frac{\partial \Omega}{\partial y}, \quad \frac{\partial \Omega}{\partial Z} = \frac{\partial \Omega}{\partial z}.$$

The components of grad Ω are therefore invariant under a translation of the coordinate axes.

(b) Consider a rotation of the axes, defined by equations (1.19), viz.

$$\begin{aligned} x &= l_{11} x' + l_{21} y' + l_{31} z', \\ y &= l_{12} x' + l_{22} y' + l_{32} z', \\ z &= l_{13} x' + l_{23} y' + l_{33} z'. \end{aligned} \tag{4.11}$$

Using the chain rule,

$$\frac{\partial \Omega}{\partial x'} = \frac{\partial x}{\partial x'} \frac{\partial \Omega}{\partial x} + \frac{\partial y}{\partial x'} \frac{\partial \Omega}{\partial y} + \frac{\partial z}{\partial x'} \frac{\partial \Omega}{\partial z},$$

$$= l_{11} \frac{\partial \Omega}{\partial x} + l_{12} \frac{\partial \Omega}{\partial y} + l_{13} \frac{\partial \Omega}{\partial z}.$$

Similarly,

$$\frac{\partial \Omega}{\partial y'} = l_{21} \frac{\partial \Omega}{\partial x} + l_{22} \frac{\partial \Omega}{\partial y} + l_{23} \frac{\partial \Omega}{\partial z},$$

and

$$\frac{\partial \Omega}{\partial z'} = l_{31} \frac{\partial \Omega}{\partial x} + l_{32} \frac{\partial \Omega}{\partial y} + l_{33} \frac{\partial \Omega}{\partial z}.$$

These equations show that under a rotation of the axes the components of grad Ω transform as components of a vector (cf. equations (2.1)), and so the verification that grad Ω is a vector is complete.

Remark. Mere existence of the first-order partial derivatives Ω_x, Ω_y, Ω_z is not sufficient to ensure that grad Ω is a vector field. For the chain rule must be used during the course of the proof, and this demands a stronger condition on Ω. We have assumed that Ω is continuously differentiable, and this (together with continuous differentiability of x, y, z as functions of the new variables, which is obviously the case by inspection of equations (4.10) and (4.11)) is certainly sufficient.

EXAMPLE 4. Find $\operatorname{grad}\Omega$ when

$$\text{(i)} \quad \Omega = x^2 + xy + y^2,$$
$$\text{(ii)} \quad \Omega = r,$$

where r denotes distance from the origin.

Solution. (i) We have

$$\frac{\partial \Omega}{\partial x} = 2x + y, \quad \frac{\partial \Omega}{\partial y} = x + 2y, \quad \frac{\partial \Omega}{\partial z} = 0.$$

Hence, from the definition (4.9),

$$\operatorname{grad}\Omega = (2x + y)\mathbf{i} + (x + 2y)\mathbf{j}.$$

(ii) We have

$$r^2 = x^2 + y^2 + z^2.$$

Hence

$$2r \frac{\partial r}{\partial x} = 2x,$$

giving

$$\frac{\partial r}{\partial x} = \frac{x}{r}.$$

Similarly

$$\frac{\partial r}{\partial y} = \frac{y}{r} \quad \text{and} \quad \frac{\partial r}{\partial z} = \frac{z}{r}.$$

Thus

$$\operatorname{grad} r = \frac{x}{r}\mathbf{i} + \frac{y}{r}\mathbf{j} + \frac{z}{r}\mathbf{k}$$
$$= (x\mathbf{i} + y\mathbf{j} + z\mathbf{k})/r$$
$$= \mathbf{r}/r = \hat{\mathbf{r}}, \tag{4.12}$$

the radial unit vector.

EXERCISES

9. If

$$\Omega = xy^2 z^3 - x^3 y^2 z,$$

find $\operatorname{grad}\Omega$ at the point $(1, -1, 1)$.

10. If

$$\Omega = x^n + y^n + z^n,$$

show that $\mathbf{r} \cdot \operatorname{grad}\Omega = n\Omega$.

11. Show that $\operatorname{grad} r^n = nr^{n-2}\mathbf{r}$.

12. If $\mathbf{a}$ is a constant vector field (i.e. a vector field of constant magnitude and direction) and $\mathbf{r}$ is the position vector, prove that

$$\operatorname{grad}(\mathbf{a}.\mathbf{r}) = \mathbf{a}.$$

4.5 Properties of gradient

The directional derivative $\partial\Omega/\partial n$. Let P be a fixed point and let P' be another point which varies in such a way that the vector $\overrightarrow{PP'}$ is always parallel to a *fixed* unit vector $\hat{\mathbf{n}}$. Assume that a scalar field Ω takes values $\Omega(P)$, $\Omega(P')$ at P, P' respectively. *The derivative of* Ω *at* P *in the direction of* $\hat{\mathbf{n}}$ is defined as

$$\frac{\partial\Omega}{\partial n} = \lim_{PP'\to 0} \frac{\Omega(P') - \Omega(P)}{PP'}, \tag{4.13}$$

whenever the limit exists. It is clear that, in general, Ω will vary at different rates as we move away from P in different directions; the directional derivative $\partial\Omega/\partial n$ measures the rate of variation in the direction of $\hat{\mathbf{n}}$.

We now have the following important property of $\operatorname{grad}\Omega$:

$$\hat{\mathbf{n}}.\operatorname{grad}\Omega = \frac{\partial\Omega}{\partial n}. \tag{4.14}$$

The proof is quite simple. Choose axes $Oxyz$ such that the directions of the unit vectors $\mathbf{i}$ and $\hat{\mathbf{n}}$ coincide. Then

$$\hat{\mathbf{n}}.\operatorname{grad}\Omega = \mathbf{i}.\operatorname{grad}\Omega = \frac{\partial\Omega}{\partial x}$$

which is the rate of change of Ω in the x-direction, that is to say the direction of $\hat{\mathbf{n}}$.

Geometrical interpretation of $\operatorname{grad}\Omega$. Let θ denote the angle between the vector $\operatorname{grad}\Omega$ and the unit vector $\hat{\mathbf{n}}$. Then

$$\frac{\partial\Omega}{\partial n} = \hat{\mathbf{n}}.\operatorname{grad}\Omega = |\operatorname{grad}\Omega|\cos\theta. \tag{4.15}$$

Suppose that θ is allowed to vary, and consider a point where $\operatorname{grad}\Omega \neq \mathbf{0}$. At this point, (4.15) shows that the greatest value of $\partial\Omega/\partial n$ will occur when $\theta = 0$; i.e. when $\hat{\mathbf{n}}$ and $\operatorname{grad}\Omega$ have the same direction. It follows that, *at a point where* $\operatorname{grad}\Omega \neq \mathbf{0}$, *the vector* $\operatorname{grad}\Omega$ *points in the direction in which* Ω *increases most rapidly, and* $|\operatorname{grad}\Omega|$ *is the rate of change of* Ω *in this direction. The gradient of a scalar field thus describes completely the manner in which the field varies.*

A point where $\text{grad}\,\Omega = \mathbf{0}$ is called a stationary point.

Level surfaces. Consider the equation

$$\Omega(x, y, z) = \lambda, \tag{4.16}$$

where λ is a parameter. For each fixed value of λ the equation represents a surface, and if λ is allowed to take a variety of different values we obtain a family of surfaces.

For example, for any fixed positive value of λ, the equation

$$x^2 + y^2 + z^2 = \lambda \tag{4.17}$$

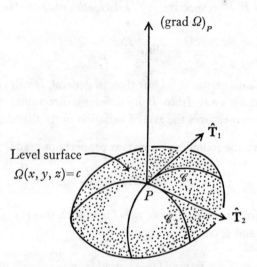

FIG. 35. The vector $(\text{grad}\,\Omega)_P$ is normal to the level surface of Ω which passes through P

represents a sphere with centre at the origin and of radius $\sqrt{\lambda}$. If λ varies, the equation represents a family of concentric spheres.

The family of surfaces (4.16) are called the *iso-surfaces* or *level surfaces* of Ω. On each level surface Ω is constant. A particular example of a set of level surfaces are the surfaces of constant atmospheric pressure. Their intersections with the earth's surface are shown on weather maps (and usually referred to as 'isobars', i.e. lines of constant barometric pressure).

An important connection between the gradient of a scalar field, and the level surfaces of the field is the following. Let P be any point on the level surface

$$\Omega(x, y, z) = c, \tag{4.18}$$

where c is a constant. Suppose that $\mathscr{C}_1$ and $\mathscr{C}_2$ are two curves through P which lie on the surface, and that $\hat{\mathbf{T}}_1$ and $\hat{\mathbf{T}}_2$ are their unit tangents at P.

Assume that grad Ω does not vanish at P, and denote its value at this point by $(\text{grad}\,\Omega)_P$. Then the vector $(\text{grad}\,\Omega)_P$ is directed along the normal to the surface at P (Fig. 35).

Proof. Let the intrinsic equation of $\mathscr{C}_1$ be

$$\mathbf{r} = \mathbf{r}(s) = (x(s), y(s), z(s)). \tag{4.19}$$

Since the curve lies on the surface (4.18), the coordinates of any point on the curve must satisfy (4.18), and so

$$\Omega(x(s), y(s), z(s)) = c.$$

Differentiating this equation with respect to s, using the chain rule, we have

$$\frac{\partial \Omega}{\partial x}\frac{dx}{ds} + \frac{\partial \Omega}{\partial y}\frac{dy}{ds} + \frac{\partial \Omega}{\partial z}\frac{dz}{ds} = 0. \tag{4.20}$$

Now

$$\text{grad}\,\Omega = \left(\frac{\partial \Omega}{\partial x}, \frac{\partial \Omega}{\partial y}, \frac{\partial \Omega}{\partial z} \right),$$

and the unit tangent to the curve (4.19) is

$$\hat{\mathbf{T}} = \left(\frac{dx}{ds}, \frac{dy}{ds}, \frac{dz}{ds} \right).$$

Thus (4.20) reduces to

$$\hat{\mathbf{T}}.\text{grad}\,\Omega = 0,$$

and hence, at P,

$$\hat{\mathbf{T}}_1.(\text{grad}\,\Omega)_P = 0.$$

Since $(\text{grad}\,\Omega)_P \neq \mathbf{0}$, the unit vector $\hat{\mathbf{T}}_1$ and the vector $(\text{grad}\,\Omega)_P$ are perpendicular. Also, by repeating the above argument, $(\text{grad}\,\Omega)_P$ is perpendicular to the unit tangent $\hat{\mathbf{T}}_2$ to $\mathscr{C}_2$. Now the unit tangents to $\mathscr{C}_1$, $\mathscr{C}_2$ are also tangents to the surface (since the curves lie on the surface). Hence, since the vector $(\text{grad}\,\Omega)_P$ is perpendicular to both these unit tangents, it is directed along the normal to the surface.

EXAMPLE 5. Let

$$\Omega(x, y, z) = x^2 + y^2 + z^2 = a^2.$$

Then

$$\text{grad}\,\Omega = (2x, 2y, 2z).$$

Thus, the vector

$$(2x_0, 2y_0, 2z_0)$$

is normal to the level surface through the point $P(x_0, y_0, z_0)$. Reference to Fig. 36, and the observation that

$$(x_0, y_0, z_0) = \overrightarrow{OP}$$

confirm this conclusion.

Taylor's expansion. Let $P(x,y,z)$ and $Q(x+\delta x, y+\delta y, z+\delta z)$ be neighbouring points in an open region $\mathscr{R}$, and put

$$\overrightarrow{PQ} = \delta \mathbf{r} = (\delta x, \delta y, \delta z). \tag{4.21}$$

If the scalar field $\Omega(x,y,z)$ is continuously differentiable in $\mathscr{R}$, it can be shown (see for example the references given in § 4.2) that

$$\Omega(x+\delta x, y+\delta y, z+\delta z) = \Omega(x,y,z) + \delta x \frac{\partial \Omega}{\partial x} + \delta y \frac{\partial \Omega}{\partial y} + \delta z \frac{\partial \Omega}{\partial z} + \epsilon |\delta \mathbf{r}|,$$

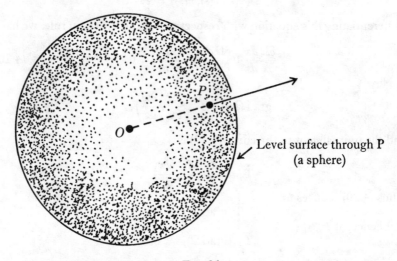

Level surface through P
(a sphere)

FIG. 36

where $\epsilon \to 0$ as $|\delta \mathbf{r}| \to 0$, and $\partial \Omega/\partial x$, $\partial \Omega/\partial y$, $\partial \Omega/\partial z$ are evaluated at $P(x,y,z)$. The notation introduced in this section enables us to express this in a more concise form, viz.

$$\Omega_Q = \Omega_P + \delta \mathbf{r} \cdot (\operatorname{grad} \Omega)_P + \epsilon |\delta \mathbf{r}|. \tag{4.22}$$

where the suffixes P, Q indicate that the quantity concerned is to be evaluated at P, Q respectively.

An explicit expression for ϵ can be given if it is assumed that all second-order derivatives of Ω exist and are continuous; the expression for Ω_Q is then the Taylor expansion of Ω, to second-order terms. However, for many purposes it is sufficient to know that $\epsilon \to 0$ as $|\delta \mathbf{r}| \to 0$.

We may note that, to a first-order approximation, (4.22) implies that the difference $\delta \Omega$ in the values of Ω at P, Q is given by the formula

$$\Omega_Q - \Omega_P = \delta \Omega \approx \delta \mathbf{r} \cdot (\operatorname{grad} \Omega)_P. \tag{4.23}$$

EXERCISES

13. Let a_1, a_2 be constants, and let Ω_1, Ω_2 be continuously differentiable scalar fields. Prove from the definition of gradient that

$$\operatorname{grad}(a_1\Omega_1 + a_1\Omega_2) \equiv a_1\operatorname{grad}\Omega_1 + a_2\operatorname{grad}\Omega_2.$$

[N.B. This is an important property of grad, showing that it is a *linear operator* – cf. § 4.7.]

14. Find the derivative of the scalar field $\Omega = x^2yz + 4xz^2$ in the direction of the vector $(2,-1,-1)$ at the point $P(1,-2,-1)$. [*Hint*. Write down the unit vector $\hat{\mathbf{n}}$ which has the same direction as the given vector, and evaluate $\hat{\mathbf{n}}.\operatorname{grad}\Omega$ at the point P.]

15. Relative to axes $Oxyz$, the temperature in a certain medium is given by

$$T = T_0(1 + ax + by)\,e^{cz},$$

where a, b, c, and T_0 (>0) are constants. At the origin O, find the direction in which the temperature changes most rapidly.

16. Find direction ratios of the normal to the ellipsoid

$$x^2/a^2 + y^2/b^2 + z^2/c^2 = \text{constant},$$

at any point; where a, b, c are constants.

4.6 The divergence and curl of a vector field

Suppose that the components of the vector field $\mathbf{F} = (F_1, F_2, F_3)$ are continuously differentiable functions of the coordinates x, y, z. Then the *divergence* of the vector field is defined as

$$\operatorname{div}\mathbf{F} = \frac{\partial F_1}{\partial x} + \frac{\partial F_2}{\partial y} + \frac{\partial F_3}{\partial z}; \tag{4.24}$$

and we define the *curl* of the vector field as

$$\operatorname{curl}\mathbf{F} = \left(\frac{\partial F_3}{\partial y} - \frac{\partial F_2}{\partial z}\right)\mathbf{i} + \left(\frac{\partial F_1}{\partial z} - \frac{\partial F_3}{\partial x}\right)\mathbf{j} + \left(\frac{\partial F_2}{\partial x} - \frac{\partial F_1}{\partial y}\right)\mathbf{k}. \tag{4.25}$$

The formula for $\operatorname{curl}\mathbf{F}$ can also be conveniently expressed in the symbolic determinantal form

$$\operatorname{curl}\mathbf{F} = \begin{vmatrix} \mathbf{i} & \mathbf{j} & \mathbf{k} \\ \dfrac{\partial}{\partial x} & \dfrac{\partial}{\partial y} & \dfrac{\partial}{\partial z} \\ F_1 & F_2 & F_3 \end{vmatrix}. \tag{4.26}$$

It is clear that $\operatorname{div}\mathbf{F}$ is a scalar field; it also has the fundamental property that it is invariant under a translation or a rotation of the coordinate axes. On the other hand, *curl* F *is a vector field*, as the notation used anticipates. The ground covered in Chapter 2 enables these statements to be proved easily, but it is convenient to defer the proofs until operator notation is introduced in the next section.

It was shown in § 4.5 that the geometrical interpretation of grad Ω is that it measures the variation of the scalar field Ω. The divergence and curl of a vector field cannot be interpreted in such simple terms, and discussion of the significance of these concepts is left for a later stage (Chapter 6). However, we may note that divergence and curl play an important role in several branches of applied mathematics; e.g. hydrodynamics, elasticity, and electromagnetism.

EXAMPLE 6. Find the divergence of the following vector fields:

$$\text{(i)} \quad \mathbf{F} = (x^2, 3y, x^3),$$
$$\text{(ii)} \quad \mathbf{F} = \mathbf{r}, \quad \text{where } \mathbf{r} \text{ is the position vector.}$$

Solution

(i) Using (4·24),
$$\operatorname{div} \mathbf{F} = \frac{\partial(x^2)}{\partial x} + \frac{\partial(3y)}{\partial y} + \frac{\partial(x^3)}{\partial z}$$
$$= 2x + 3.$$

(ii) Since $\mathbf{r} = (x, y, z)$,
$$\operatorname{div} \mathbf{r} = 3. \tag{4.27}$$

EXAMPLE 7. Find curl F for the vector field

$$\mathbf{F} = (z, x, y),$$

and show that curl curl F $= \mathbf{0}$.

Solution. We have

$$\operatorname{curl} \mathbf{F} = \begin{vmatrix} \mathbf{i} & \mathbf{j} & \mathbf{k} \\ \dfrac{\partial}{\partial x} & \dfrac{\partial}{\partial y} & \dfrac{\partial}{\partial z} \\ z & x & y \end{vmatrix}$$

$$= (1-0)\mathbf{i} + (1-0)\mathbf{j} + (1-0)\mathbf{k}$$
$$= \mathbf{i} + \mathbf{j} + \mathbf{k}$$
$$= (1, 1, 1).$$

Also

$$\operatorname{curl} \operatorname{curl} \mathbf{F} = \begin{vmatrix} \mathbf{i} & \mathbf{j} & \mathbf{k} \\ \dfrac{\partial}{\partial x} & \dfrac{\partial}{\partial y} & \dfrac{\partial}{\partial z} \\ 1 & 1 & 1 \end{vmatrix}$$

$$= \mathbf{0}.$$

EXERCISES

17. If a_1, a_2 are constants and $\mathbf{F}_1$, $\mathbf{F}_2$ are continuously differentiable vector fields, prove from the definitions of divergence and curl that

$$\operatorname{div}(a_1\mathbf{F}_1 + a_2\mathbf{F}_2) \equiv a_1\operatorname{div}\mathbf{F}_1 + a_2\operatorname{div}\mathbf{F}_2,$$

and

$$\operatorname{curl}(a_1\mathbf{F}_1 + a_2\mathbf{F}_2) \equiv a_1\operatorname{curl}\mathbf{F}_1 + a_2\operatorname{curl}\mathbf{F}_2.$$

[*N.B.* These are important properties, showing that, like grad (see p. 85), div and curl are linear operators (cf. § 4.7, which follows).]

18. Find the divergence and curl of the vector field $\mathbf{F} = (xy, yz, 0)$ at the point $(1, 1, 1)$. Also evaluate $\operatorname{grad}(\operatorname{div}\mathbf{F})$.

19. If $\Omega = x + y^2 + z^3$, find $\operatorname{div}(\operatorname{grad}\Omega)$ and $\operatorname{curl}(\operatorname{grad}\Omega)$.

20. If $\mathbf{a}$ is a constant vector field and $\mathbf{r}$ is the position vector, show that

$$\operatorname{curl}(\mathbf{a} \times \mathbf{r}) = 2\mathbf{a}.$$

21. The point $A(a, b, c)$ is fixed, and $P(x, y, z)$ is a variable point. Show that

$$\operatorname{div}\overrightarrow{AP} = 3, \quad \operatorname{curl}\overrightarrow{AP} = \mathbf{0}.$$

22. At all points, a vector field $\mathbf{F}$ is parallel to the xy-plane. If the components of $\mathbf{F}$ are functions of x and y only, prove that $\operatorname{curl}\operatorname{curl}\mathbf{F}$ is parallel to the xy-plane.

4.7 The del-operator

The reader may be familiar with the concept of operators in connection with ordinary differential equations. Thus, the differential equation

$$\frac{d^2y}{dx^2} + 2\frac{dy}{dx} + 3y = 0$$

is sometimes written in the form

$$\left(\frac{d^2}{dx^2} + 2\frac{d}{dx} + 3\right)y = 0,$$

and expressed as

$$(D^2 + 2D + 3)y = 0,$$

where $D \equiv d/dx$. In this example, D is an *operator*. When it acts upon the function $y(x)$ it gives the derivative dy/dx; and when D^2 acts upon $y(x)$ it gives the second derivative d^2y/dx^2.

The D-operator obeys some, but not all, of the rules of ordinary algebra. For example

$$D(Dy) \equiv D^2y$$

7

and

$$(D^2 + 2D + 3)y \equiv D^2 y + 2Dy + 3y;$$

but

$$D(xy) \not\equiv xDy \not\equiv xyD.$$

Motivated by these ideas, we now turn to operators in vector analysis.

The del-operator. The expression

$$\nabla \equiv \mathbf{i}\frac{\partial}{\partial x} + \mathbf{j}\frac{\partial}{\partial y} + \mathbf{k}\frac{\partial}{\partial z} \equiv \left(\frac{\partial}{\partial x}, \frac{\partial}{\partial y}, \frac{\partial}{\partial z}\right), \qquad (4.28)$$

is called the *del-operator*; or, more briefly, *del* or *nabla*.

Under a translation or rotation of the axes, the components $\partial/\partial x$, $\partial/\partial y$, $\partial/\partial z$ of the del-operator transform as do the components of a vector (this explains the notation used in (4.28)). The proof of this can be obtained at once from the proof in § 4.4 that $\operatorname{grad}\Omega$ is a vector field. For, if in parts (a) and (b) of the proof, Ω is omitted from every line, we obtain

$$\frac{\partial}{\partial X} = \frac{\partial}{\partial x}, \quad \frac{\partial}{\partial Y} = \frac{\partial}{\partial y}, \quad \frac{\partial}{\partial Z} = \frac{\partial}{\partial z};$$

and

$$\frac{\partial}{\partial x'} = l_{11}\frac{\partial}{\partial x} + l_{12}\frac{\partial}{\partial y} + l_{13}\frac{\partial}{\partial z},$$

together with similar expressions for $\partial/\partial y'$ and $\partial/\partial z'$. In this respect, then, the del-operator behaves as a vector, and it is often called a vector operator. The ideas associated with the D-operator, discussed above, can now be extended to the del-operator.

We define

$$\nabla\Omega = \operatorname{grad}\Omega. \qquad (4.29)$$

The left-hand side is

$$\left(\mathbf{i}\frac{\partial}{\partial x} + \mathbf{j}\frac{\partial}{\partial y} + \mathbf{k}\frac{\partial}{\partial z}\right)\Omega,$$

and it is natural to interpret this as

$$\mathbf{i}\frac{\partial\Omega}{\partial x} + \mathbf{j}\frac{\partial\Omega}{\partial y} + \mathbf{k}\frac{\partial\Omega}{\partial z} = \operatorname{grad}\Omega.$$

It is also natural to define

$$\nabla\,.\,\mathbf{F} = \operatorname{div}\mathbf{F} \qquad (4.30)$$

and

$$\nabla\times\mathbf{F} = \operatorname{curl}\mathbf{F}. \qquad (4.31)$$

For, expanding the scalar product formally,

$$\mathbf{\nabla}.\mathbf{F} = \left(\frac{\partial}{\partial x}, \frac{\partial}{\partial y}, \frac{\partial}{\partial z}\right).(F_1, F_2, F_3)$$

$$= \frac{\partial F_1}{\partial x} + \frac{\partial F_2}{\partial y} + \frac{\partial F_3}{\partial z}$$

$$= \operatorname{div}\mathbf{F};$$

and likewise we have

$$\mathbf{\nabla}\times\mathbf{F} = \left(\frac{\partial}{\partial x}, \frac{\partial}{\partial y}, \frac{\partial}{\partial z}\right)\times(F_1, F_2, F_3)$$

$$= \begin{vmatrix} \mathbf{i} & \mathbf{j} & \mathbf{k} \\ \dfrac{\partial}{\partial x} & \dfrac{\partial}{\partial y} & \dfrac{\partial}{\partial z} \\ F_1 & F_2 & F_3 \end{vmatrix}$$

$$= \operatorname{curl}\mathbf{F}.$$

Observe that, as with the D-operator, the components of $\mathbf{\nabla}$ act only upon functions on their right.

We are now in a position to prove two results left over from the previous section, viz.

(a) *The scalar field* $\operatorname{div}\mathbf{F}$ *is invariant under a translation or rotation of the coordinate axes*;

(b) $\operatorname{curl}\mathbf{F}$ *is a vector field.*

Proofs.

(a) Writing

$$\operatorname{div}\mathbf{F} = \mathbf{\nabla}.\mathbf{F},$$

the invariance under a translation of the axes follows easily. For, it has been seen that the components of $\mathbf{\nabla}$ are invariant under a translation of the coordinate axes, and since the components of the *vector* $\mathbf{F}$ are also invariant under such a transformation, $\mathbf{\nabla}.\mathbf{F}$ is invariant.

To establish the invariance of $\mathbf{\nabla}.\mathbf{F}$ under a rotation of the axes, it is convenient to denote the coordinates by x_1, x_2, x_3 instead of x, y, z. With this notation

$$\mathbf{\nabla}.\mathbf{F} = \frac{\partial F_1}{\partial x_1} + \frac{\partial F_2}{\partial x_2} + \frac{\partial F_3}{\partial x_3} = \frac{\partial F_i}{\partial x_i},$$

using the summation convention.

Consider the rotation to new axes $Ox_1'x_2'x_3'$ defined by equations (1.25).

Using primes to denote components relative to the new axes, the vector transformation law (equation (2.2)) gives

$$\frac{\partial}{\partial x'_j} = l_{ji} \frac{\partial}{\partial x_i},$$

and

$$F'_j = l_{jk} F_k.$$

Hence, since the coefficients l_{jk} are independent of x_1, x_2, x_3,

$$\frac{\partial F'_j}{\partial x'_j} = l_{ji} l_{jk} \frac{\partial F_k}{\partial x_i}.$$

Using the orthonormality conditions

$$l_{ji} l_{jk} = \delta_{ik}$$

(equation (1.29)), it follows that

$$\frac{\partial F'_j}{\partial x'_j} = \delta_{ik} \frac{\partial F_k}{\partial x_i} = \frac{\partial F_i}{\partial x_i};$$

i.e.

$$\frac{\partial F'_1}{\partial x'_1} + \frac{\partial F'_2}{\partial x'_2} + \frac{\partial F'_3}{\partial x'_3} = \frac{\partial F_1}{\partial x_1} + \frac{\partial F_2}{\partial x_2} + \frac{\partial F_3}{\partial x_3}.$$

The left-hand side is $\operatorname{div} \mathbf{F}$ referred to the new axes, and the right-hand side is $\operatorname{div} \mathbf{F}$ referred to the original axes; the invariance of $\operatorname{div} \mathbf{F}$ under a rotation of the axes is thus established.

(b) If the reader refers to § 2.6, p. 32, he will see that the proof of the invariance of $\nabla . \mathbf{F}$ follows closely the proof that the scalar product $\mathbf{a} . \mathbf{b}$ is a scalar invariant. Similarly, the proof that $\nabla \times \mathbf{F}$ is a vector field follows closely the proof in § 2.7, p. 36, that the vector product $\mathbf{a} \times \mathbf{b}$ is a vector. The essential features are similar because the operator ∇ transforms under a translation or rotation of the axes as a vector. It is left as an exercise for the reader to verify this.

We have already remarked, in Exercise 13, p. 85, and Exercise 17, p. 87, that grad, div, and curl are all linear operators; and we can now express these same facts concisely in the single statement that del is a linear operator.

The differential operator $D \equiv d/dx$, to which we have already referred, possesses the characteristic property of linearity:

$$D(a_1 y_1 + a_2 y_2) = a_1 D y_1 + a_2 D y_2$$

for any constants a_1, a_2; and in the same way, the partial differential operator $\partial/\partial x$ may be said to be linear, since

$$\frac{\partial}{\partial x}(a_1 f_1 + a_2 f_2) = a_1 \frac{\partial f_1}{\partial x} + a_2 \frac{\partial f_2}{\partial x}$$

for any functions f_1, f_2 that can be differentiated partially with respect to x, and any constants a_1, a_2.

Now

$$\nabla \equiv \mathbf{i}\frac{\partial}{\partial x} + \mathbf{j}\frac{\partial}{\partial y} + \mathbf{k}\frac{\partial}{\partial z}$$

is a linear combination of such partial differential operators, and we may therefore expect it also to behave as a linear operator. As the reader can easily verify, ∇ does indeed behave in this way in all three of its possible applications: $\nabla\Omega$, $\nabla \cdot \mathbf{F}$, and $\nabla \times \mathbf{F}$, as is asserted in Exercises 13 and 17.

EXERCISES

23. Express in operator notation: (i) $\text{grad}(\text{div}\,\mathbf{F})$, (ii) $\text{div}(\text{grad}\,\Omega)$, (iii) $\text{div}(\text{curl}\,\mathbf{F})$, (iv) $\text{curl}(\text{grad}\,\Omega)$, (v) $\text{curl curl}\,\mathbf{F}$.

24. Prove that $(\mathbf{F} \times \nabla) \times \mathbf{G}$ is a vector field.

4.8 Scalar invariant operators

Let $\mathscr{D}$ denote a linear partial differential operator which involves only the rectangular cartesian coordinates x, y, z as independent variables. For example, we might have

$$\mathscr{D} = \frac{\partial}{\partial x} + \frac{\partial}{\partial y} + 2\frac{\partial}{\partial z} \qquad \text{or} \qquad \mathscr{D} = \frac{\partial^2}{\partial x^2} + \frac{\partial}{\partial y} + \frac{\partial^2}{\partial y\,\partial z}.$$

The operator $\mathscr{D}$ is called a *scalar invariant operator* if its form is unchanged under a translation or rotation of the coordinate axes. Thus, for example, if the first of the operators above were invariant (it is not, in fact), then upon changing to new axes $0'x'y'z'$ it would become

$$\mathscr{D}' = \frac{\partial}{\partial x'} + \frac{\partial}{\partial y'} + 2\frac{\partial}{\partial z'}.$$

The following result concerning scalar invariant operators will be required.

Let $\mathscr{D}$ be a scalar invariant operator, and define its operation on a vector field $\mathbf{F}$ by

$$\mathscr{D}\mathbf{F} = \mathscr{D}(F_1, F_2, F_3) = (\mathscr{D}F_1, \mathscr{D}F_2, \mathscr{D}F_3). \tag{4.32}$$

Then $\mathscr{D}\mathbf{F}$ is a vector field.

Proof. Under a translation of the axes, the components of $\mathscr{D}\mathbf{F}$ are unchanged, because both $\mathscr{D}$ and the components of $\mathbf{F}$ are unchanged.

Under a rotation to new axes $Ox'y'z'$, whose positions relative to the original axes $Oxyz$ are defined by the transformation matrix (1.12), we have

$$F_j' = l_{jk}F_k,$$

where, as usual, primes denote components relative to the new axes. Thus, using the property of invariance of the form of $\mathcal{D}$, and also the linearity of $\mathcal{D}$,

$$\mathcal{D}' F_j' = \mathcal{D}(l_{jk} F_k) = l_{jk} \mathcal{D} F_k,$$

showing that the components of $\mathcal{D}\mathbf{F}$ transform according to the vector law under a rotation of the axes. It follows that $\mathcal{D}\mathbf{F}$ is a vector field.

The Laplacian operator ∇^2. The most important of the scalar invariant operators is the *Laplacian*, defined as

$$\nabla^2 \equiv \frac{\partial^2}{\partial x^2} + \frac{\partial^2}{\partial y^2} + \frac{\partial^2}{\partial z^2}. \tag{4.33}$$

Alternatively, using rectangular coordinates x_1, x_2, x_3 and the summation convention,

$$\nabla^2 \equiv \frac{\partial^2}{\partial x_1^2} + \frac{\partial^2}{\partial x_2^2} + \frac{\partial^2}{\partial x_3^2} \equiv \frac{\partial}{\partial x_i} \frac{\partial}{\partial x_i}. \tag{4.34}$$

Formally, the Laplacian operator is the 'square' of the del-operator. This connection is easily understood by observing that, formally,

$$\mathbf{\nabla . \nabla} = \left(\frac{\partial}{\partial x}, \frac{\partial}{\partial y}, \frac{\partial}{\partial z} \right) . \left(\frac{\partial}{\partial x}, \frac{\partial}{\partial y}, \frac{\partial}{\partial z} \right) = \frac{\partial^2}{\partial x^2} + \frac{\partial^2}{\partial y^2} + \frac{\partial^2}{\partial z^2}.$$

By analogy with the relation $\mathbf{a . a} = a^2$, it is thus natural to use the notation $\mathbf{\nabla . \nabla} = \nabla^2$.

The invariance of the Laplacian operator follows from the fact that the del-operator transforms as a vector. Thus, under a translation of the axes, $\partial/\partial x_i$ is invariant and hence so also is ∇^2. Under a rotation to coordinate axes $Ox_1'x_2'x_3'$, we have

$$\frac{\partial}{\partial x_j'} = l_{ji} \frac{\partial}{\partial x_i}$$

(cf. the proof of the invariance of $\mathrm{div}\,\mathbf{F}$ in § 4.7, p. 89). Thus

$$\frac{\partial}{\partial x_j'} \frac{\partial}{\partial x_j'} = \left(l_{ji} \frac{\partial}{\partial x_i} \right) \left(l_{jk} \frac{\partial}{\partial x_k} \right)$$

$$= \delta_{ik} \frac{\partial}{\partial x_i} \frac{\partial}{\partial x_k}$$

$$= \frac{\partial}{\partial x_i} \frac{\partial}{\partial x_i},$$

which shows that ∇^2 is also invariant under a rotation of the axes.

The Laplacian operator can act upon a scalar field Ω or on a vector field $\mathbf{F}$. The following important identities hold:

$$\nabla^2 \Omega \equiv \text{div}(\text{grad}\,\Omega); \tag{4.35}$$

$$\nabla^2 \mathbf{F} \equiv \text{grad}(\text{div}\,\mathbf{F}) - \text{curl curl}\,\mathbf{F}. \tag{4.36}$$

These results are proved by expanding the right-hand sides. Thus,

$$\text{div}(\text{grad}\,\Omega) \equiv \text{div}\left(\mathbf{i}\frac{\partial\Omega}{\partial x} + \mathbf{j}\frac{\partial\Omega}{\partial y} + \mathbf{k}\frac{\partial\Omega}{\partial z}\right)$$

$$\equiv \frac{\partial}{\partial x}\left(\frac{\partial\Omega}{\partial x}\right) + \frac{\partial}{\partial y}\left(\frac{\partial\Omega}{\partial y}\right) + \frac{\partial}{\partial z}\left(\frac{\partial\Omega}{\partial z}\right)$$

$$\equiv \nabla^2\Omega,$$

which proves (4.35). If $\mathbf{F} = (F_1, F_2, F_3)$, the x-component of $\text{grad}(\text{div}\,\mathbf{F})$ is

$$\frac{\partial}{\partial x}\left(\frac{\partial F_1}{\partial x} + \frac{\partial F_2}{\partial y} + \frac{\partial F_3}{\partial z}\right) \equiv \frac{\partial^2 F_1}{\partial x^2} + \frac{\partial^2 F_2}{\partial x\,\partial y} + \frac{\partial^2 F_3}{\partial x\,\partial z}.$$

Also,

$$\text{curl curl}\,\mathbf{F} \equiv \begin{vmatrix} \mathbf{i} & \mathbf{j} & \mathbf{k} \\ \dfrac{\partial}{\partial x} & \dfrac{\partial}{\partial y} & \dfrac{\partial}{\partial z} \\ \dfrac{\partial F_3}{\partial y} - \dfrac{\partial F_2}{\partial z} & \dfrac{\partial F_1}{\partial z} - \dfrac{\partial F_3}{\partial x} & \dfrac{\partial F_2}{\partial x} - \dfrac{\partial F_1}{\partial y} \end{vmatrix}.$$

Thus the x-component of $\text{curl curl}\,\mathbf{F}$ is

$$\frac{\partial}{\partial y}\left(\frac{\partial F_2}{\partial x} - \frac{\partial F_1}{\partial y}\right) - \frac{\partial}{\partial z}\left(\frac{\partial F_1}{\partial z} - \frac{\partial F_3}{\partial x}\right) \equiv \frac{\partial^2 F_2}{\partial y\,\partial x} - \frac{\partial^2 F_1}{\partial y^2} - \frac{\partial^2 F_1}{\partial z^2} + \frac{\partial^2 F_3}{\partial z\,\partial x}.$$

Hence, assuming that the order of differentiation in the mixed derivatives can be inverted (see the theorem on p. 76), the x-component of the right-hand side of (4.36) is

$$\frac{\partial^2 F_1}{\partial x^2} + \frac{\partial^2 F_1}{\partial y^2} + \frac{\partial^2 F_1}{\partial z^2} \equiv \nabla^2 F_1.$$

The y- and z-components of the right-hand side of (4.36) can be similarly evaluated, giving finally

$$\text{grad}(\text{div}\,\mathbf{F}) - \text{curl curl}\,\mathbf{F} \equiv (\nabla^2 F_1, \nabla^2 F_2, \nabla^2 F_3) \equiv \nabla^2\mathbf{F},$$

as required.

The Laplacian operator occurs in several fundamental differential equations governing physical phenomena. The simplest of these differential equations is *Laplace's equation*, viz.

$$\nabla^2\Omega = 0. \tag{4.37}$$

The operator $\mathbf{F}.\boldsymbol{\nabla}$. This is a useful scalar invariant operator, defined as

$$\mathbf{F}.\boldsymbol{\nabla} \equiv F_1\frac{\partial}{\partial x}+F_2\frac{\partial}{\partial y}+F_3\frac{\partial}{\partial z}. \tag{4.38}$$

The proof of invariance is straightforward and is left to the reader.

The operator can act upon a scalar field Ω, giving

$$(\mathbf{F}.\boldsymbol{\nabla})\Omega = F_1\frac{\partial\Omega}{\partial x}+F_2\frac{\partial\Omega}{\partial y}+F_3\frac{\partial\Omega}{\partial z}. \tag{4.39}$$

Since the right-hand side of (4.39) is $\mathbf{F}.(\boldsymbol{\nabla}\Omega)$, no ambiguity can arise by writing

$$(\mathbf{F}.\boldsymbol{\nabla})\Omega = \mathbf{F}.\boldsymbol{\nabla}\Omega = \mathbf{F}.\text{grad}\,\Omega. \tag{4.40}$$

The operator $\mathbf{F}.\boldsymbol{\nabla}$ can also act upon a vector field $\mathbf{G}=(G_1, G_2, G_3)$, giving in rectangular cartesian coordinates

$$(\mathbf{F}.\boldsymbol{\nabla})\mathbf{G} = (\mathbf{F}.\boldsymbol{\nabla}G_1, \mathbf{F}.\boldsymbol{\nabla}G_2, \mathbf{F}.\boldsymbol{\nabla}G_3). \tag{4.41}$$

The result proved at the beginning of this section shows that $(\mathbf{F}.\boldsymbol{\nabla})\mathbf{G}$ is a vector field.

N.B. Interpretation of the expressions $\nabla^2\mathbf{F}$ and $(\mathbf{F}.\boldsymbol{\nabla})\mathbf{G}$ requires a little care when orthogonal coordinates other than the rectangular cartesian system are used (see Example 14, p. 113, and Exercise 52, p. 115).

EXERCISES

25. If $\mathbf{F}_1=(x,y,z)$, $\mathbf{F}_2=(1,2,3)$, $\mathbf{G}=(x^2,y^2,z^2)$, and $\Omega=xyz$, evaluate:

(i) $(\mathbf{F}_1.\boldsymbol{\nabla})\mathbf{G}$,　　(ii) $(\mathbf{F}_2.\boldsymbol{\nabla})\mathbf{G}$,　　(iii) $(\mathbf{F}_1.\boldsymbol{\nabla})\mathbf{F}_2$,
(iv) $\mathbf{F}_1.\boldsymbol{\nabla}\Omega$,　　(v) $\nabla^2\mathbf{G}$,　　(vi) $\boldsymbol{\nabla}\times\{(\mathbf{F}_2.\boldsymbol{\nabla})\mathbf{G}\}$.

26. If $\mathbf{F}=(xy, yz, zx)$ and $\mathbf{a}=(1,2,3)$, verify that:

(i) $\boldsymbol{\nabla}\times(\boldsymbol{\nabla}\times\mathbf{F}) \equiv \boldsymbol{\nabla}(\boldsymbol{\nabla}.\mathbf{F})-\nabla^2\mathbf{F}$,
(ii) $\boldsymbol{\nabla}\times(\mathbf{F}\times\mathbf{a}) \equiv \mathbf{F}(\boldsymbol{\nabla}.\mathbf{a})-\mathbf{a}(\boldsymbol{\nabla}.\mathbf{F})+(\mathbf{a}.\boldsymbol{\nabla})\mathbf{F}-(\mathbf{F}.\boldsymbol{\nabla})\mathbf{a}$.

4.9　Useful identities

Alternative expressions for $\text{grad}\,\phi$, *div* $\mathbf{F}$ *and curl* $\mathbf{F}$. It is convenient to introduce here new expressions for $\text{grad}\,\phi$, $\text{div}\,\mathbf{F}$ and $\text{curl}\,\mathbf{F}$ which make the proofs of some of the identities given later less laborious.

Take rectangular cartesian axes Ox_1, Ox_2, Ox_3, and let $\mathbf{e}_1$, $\mathbf{e}_2$, $\mathbf{e}_3$ denote the unit vectors along the respective axes. Then, *using the summation convention*,

(a)　　　　　　　$$\text{grad}\,\Omega = \mathbf{e}_i\frac{\partial\Omega}{\partial x_i}, \tag{4.42}$$

(b)
$$\operatorname{div} \mathbf{F} = \mathbf{e}_i \cdot \frac{\partial \mathbf{F}}{\partial x_i}, \tag{4.43}$$

(c)
$$\operatorname{curl} \mathbf{F} = \mathbf{e}_i \times \frac{\partial \mathbf{F}}{\partial x_i}, \tag{4.44}$$

where
$$\frac{\partial \mathbf{F}}{\partial x_i} = \left(\frac{\partial F_1}{\partial x_i}, \frac{\partial F_2}{\partial x_i}, \frac{\partial F_3}{\partial x_i} \right), \quad i = 1, 2, 3. \tag{4.45}$$

Proof. (a) Expression (4.42) follows at once from the definition of $\operatorname{grad} \Omega$.
(b) Since $\mathbf{F} = F_1 \mathbf{e}_1 + F_2 \mathbf{e}_2 + F_3 \mathbf{e}_3 = F_j \mathbf{e}_j$, and the vectors $\mathbf{e}_j$ are constant,

$$\mathbf{e}_i \cdot \frac{\partial \mathbf{F}}{\partial x_i} = \mathbf{e}_i \cdot \mathbf{e}_j \frac{\partial F_j}{\partial x_i}$$

$$= \delta_{ij} \frac{\partial F_j}{\partial x_i},$$

using the fact that $\mathbf{e}_1$, $\mathbf{e}_2$, $\mathbf{e}_3$ form an orthonormal triad. Thus

$$\mathbf{e}_i \cdot \frac{\partial \mathbf{F}}{\partial x_i} = \frac{\partial F_i}{\partial x_i} = \operatorname{div} \mathbf{F}.$$

(c) We have

$$\mathbf{e}_i \times \frac{\partial \mathbf{F}}{\partial x_i} = \mathbf{e}_i \times \mathbf{e}_j \frac{\partial F_j}{\partial x_i}.$$

The terms obtained by allowing each of i, j to take the values 1, 2 are

$$\mathbf{e}_1 \times \mathbf{e}_1 \frac{\partial F_1}{\partial x_1} + \mathbf{e}_1 \times \mathbf{e}_2 \frac{\partial F_2}{\partial x_1} + \mathbf{e}_2 \times \mathbf{e}_1 \frac{\partial F_1}{\partial x_2} + \mathbf{e}_2 \times \mathbf{e}_2 \frac{\partial F_2}{\partial x_2} = \left(\frac{\partial F_2}{\partial x_1} - \frac{\partial F_1}{\partial x_2} \right) \mathbf{e}_3,$$

and this is the x_3-component of $\operatorname{curl} \mathbf{F}$. Similarly the other two possible pairs of values of i, j give the x_1- and x_2-components of $\operatorname{curl} \mathbf{F}$. Hence (4.44) follows.

Identities. If it is assumed (in accordance with our usual practice) that all necessary derivatives exist and are continuous, then the following identities hold:

(i) $\operatorname{div}(\operatorname{curl} \mathbf{F}) \equiv 0,$ (4.46)

(ii) $\operatorname{curl}(\operatorname{grad} \Omega) \equiv \mathbf{0},$ (4.47)

(iii) $\operatorname{grad}(\Omega_1 \Omega_2) \equiv \Omega_1 \operatorname{grad} \Omega_2 + \Omega_2 \operatorname{grad} \Omega_1,$ (4.48)

(iv) $\operatorname{div}(\Omega \mathbf{F}) \equiv \Omega \operatorname{div} \mathbf{F} + \mathbf{F} \cdot \operatorname{grad} \Omega,$ (4.49)

(v) $\operatorname{curl}(\Omega \mathbf{F}) \equiv \Omega \operatorname{curl} \mathbf{F} - \mathbf{F} \times \operatorname{grad} \Omega,$ (4.50)

(vi) $\operatorname{grad}(\mathbf{F}.\mathbf{G}) \equiv \mathbf{F} \times \operatorname{curl}\mathbf{G} + \mathbf{G} \times \operatorname{curl}\mathbf{F}$
$$+ (\mathbf{F}.\boldsymbol{\nabla})\mathbf{G} + (\mathbf{G}.\boldsymbol{\nabla})\mathbf{F}, \qquad (4.51)$$

(vii) $\operatorname{div}(\mathbf{F} \times \mathbf{G}) \equiv \mathbf{G}.\operatorname{curl}\mathbf{F} - \mathbf{F}.\operatorname{curl}\mathbf{G}, \qquad (4.52)$

(viii) $\operatorname{curl}(\mathbf{F} \times \mathbf{G}) \equiv \mathbf{F}\operatorname{div}\mathbf{G} - \mathbf{G}\operatorname{div}\mathbf{F} + (\mathbf{G}.\boldsymbol{\nabla})\mathbf{F} - (\mathbf{F}.\boldsymbol{\nabla})\mathbf{G}. \qquad (4.53)$

In operator notation:

(i)' $\boldsymbol{\nabla}.(\boldsymbol{\nabla} \times \mathbf{F}) \equiv 0, \qquad (4.54)$

(ii)' $\boldsymbol{\nabla} \times (\boldsymbol{\nabla}\Omega) \equiv \mathbf{0}, \qquad (4.55)$

(iii)' $\boldsymbol{\nabla}(\Omega_1\Omega_2) \equiv \Omega_1\boldsymbol{\nabla}\Omega_2 + \Omega_2\boldsymbol{\nabla}\Omega_1, \qquad (4.56)$

(iv)' $\boldsymbol{\nabla}.(\Omega\mathbf{F}) \equiv \Omega\boldsymbol{\nabla}.\mathbf{F} + \mathbf{F}.\boldsymbol{\nabla}\Omega, \qquad (4.57)$

(v)' $\boldsymbol{\nabla} \times (\Omega\mathbf{F}) \equiv \Omega\boldsymbol{\nabla} \times \mathbf{F} - \mathbf{F} \times \boldsymbol{\nabla}\Omega, \qquad (4.58)$

(vi)' $\boldsymbol{\nabla}(\mathbf{F}.\mathbf{G}) \equiv \mathbf{F} \times (\boldsymbol{\nabla} \times \mathbf{G}) + \mathbf{G} \times (\boldsymbol{\nabla} \times \mathbf{F})$
$$+ (\mathbf{F}.\boldsymbol{\nabla})\mathbf{G} + (\mathbf{G}.\boldsymbol{\nabla})\mathbf{F}, \qquad (4.59)$$

(vii)' $\boldsymbol{\nabla}.(\mathbf{F} \times \mathbf{G}) \equiv \mathbf{G}.(\boldsymbol{\nabla} \times \mathbf{F}) - \mathbf{F}.(\boldsymbol{\nabla} \times \mathbf{G}), \qquad (4.60)$

(viii)' $\boldsymbol{\nabla} \times (\mathbf{F} \times \mathbf{G}) \equiv \mathbf{F}(\boldsymbol{\nabla}.\mathbf{G}) - \mathbf{G}(\boldsymbol{\nabla}.\mathbf{F}) + (\mathbf{G}.\boldsymbol{\nabla})\mathbf{F} - (\mathbf{F}.\boldsymbol{\nabla})\mathbf{G}. \qquad (4.61)$

Proofs

(i) $\operatorname{div}(\operatorname{curl}\mathbf{F}) \equiv \dfrac{\partial}{\partial x}\left(\dfrac{\partial F_3}{\partial y} - \dfrac{\partial F_2}{\partial z}\right) + \dfrac{\partial}{\partial y}\left(\dfrac{\partial F_1}{\partial z} - \dfrac{\partial F_3}{\partial x}\right) + \dfrac{\partial}{\partial z}\left(\dfrac{\partial F_2}{\partial x} - \dfrac{\partial F_1}{\partial y}\right)$
$$\equiv 0,$$

using the assumption that the derivatives are continuous, so that the order of differentiation can be inverted.

(ii)
$$\operatorname{curl}(\operatorname{grad}\Omega) \equiv \begin{vmatrix} \mathbf{i} & \mathbf{j} & \mathbf{k} \\ \dfrac{\partial}{\partial x} & \dfrac{\partial}{\partial y} & \dfrac{\partial}{\partial z} \\ \dfrac{\partial\Omega}{\partial x} & \dfrac{\partial\Omega}{\partial y} & \dfrac{\partial\Omega}{\partial z} \end{vmatrix}$$
$$\equiv \mathbf{0},$$

because $\partial^2\Omega/\partial y\,\partial z - \partial^2\Omega/\partial z\,\partial y = 0$, with two other similar relations.

(iii) Using (4.42),
$$\operatorname{grad}(\Omega_1\Omega_2) \equiv \mathbf{e}_i\dfrac{\partial}{\partial x_i}(\Omega_1\Omega_2)$$
$$\equiv \Omega_1\mathbf{e}_i\dfrac{\partial\Omega_2}{\partial x_i} + \Omega_2\mathbf{e}_i\dfrac{\partial\Omega_1}{\partial x_i}$$
$$\equiv \Omega_1\operatorname{grad}\Omega_2 + \Omega_2\operatorname{grad}\Omega_1.$$

(iv) Using (4.43),

$$\operatorname{div}(\Omega \mathbf{F}) \equiv \mathbf{e}_i \cdot \frac{\partial}{\partial x_i}(\Omega \mathbf{F})$$

$$\equiv \Omega \mathbf{e}_i \cdot \frac{\partial \mathbf{F}}{\partial x_i} + \left(\mathbf{e}_i \frac{\partial \Omega}{\partial x_i}\right) . \mathbf{F}$$

$$\equiv \Omega \operatorname{div} \mathbf{F} + \mathbf{F} . \operatorname{grad} \Omega.$$

(v) The proof is similar to the proof of (iv) above and is left to the reader as an exercise.

(vi) Using (4.44), we have

$$\mathbf{F} \times \operatorname{curl} \mathbf{G} + \mathbf{G} \times \operatorname{curl} \mathbf{F} \equiv \mathbf{F} \times \left(\mathbf{e}_i \times \frac{\partial \mathbf{G}}{\partial x_i}\right) + \mathbf{G} \times \left(\mathbf{e}_i \times \frac{\partial \mathbf{F}}{\partial x_i}\right)$$

$$F \times (\nabla \times G) + G \times (\nabla \times F)$$

$$\equiv \left(\mathbf{F} . \frac{\partial \mathbf{G}}{\partial x_i}\right)\mathbf{e}_i - (\mathbf{F}.\mathbf{e}_i)\frac{\partial \mathbf{G}}{\partial x_i} + \left(\mathbf{G}.\frac{\partial \mathbf{F}}{\partial x_i}\right)\mathbf{e}_i$$

$$- (\mathbf{G}.\mathbf{e}_i)\frac{\partial \mathbf{F}}{\partial x_i}$$

$$\equiv \mathbf{e}_i \frac{\partial}{\partial x_i}(\mathbf{F}.\mathbf{G}) - \left(\mathbf{F}.\mathbf{e}_i\frac{\partial}{\partial x_i}\right)\mathbf{G} - \left(\mathbf{G}.\mathbf{e}_i\frac{\partial}{\partial x_i}\right)\mathbf{F}$$

$$\equiv \operatorname{grad}(\mathbf{F}.\mathbf{G}) - (\mathbf{F}.\nabla)\mathbf{G} - (\mathbf{G}.\nabla)\mathbf{F},$$

which proves identity (vi).

(vii) Using (4.43),

$$\operatorname{div}(\mathbf{F} \times \mathbf{G}) \equiv \mathbf{e}_i . \frac{\partial}{\partial x_i}(\mathbf{F} \times \mathbf{G})$$

$$\equiv \mathbf{e}_i . \left(\frac{\partial \mathbf{F}}{\partial x_i} \times \mathbf{G}\right) + \mathbf{e}_i . \left(\mathbf{F} \times \frac{\partial \mathbf{G}}{\partial x_i}\right)$$

$$\equiv \left(\mathbf{e}_i \times \frac{\partial \mathbf{F}}{\partial x_i}\right) . \mathbf{G} - \left(\mathbf{e}_i \times \frac{\partial \mathbf{G}}{\partial x_i}\right) . \mathbf{F}$$

$$\equiv \mathbf{G} . \operatorname{curl} \mathbf{F} - \mathbf{F} . \operatorname{curl} \mathbf{G},$$

using (4.44).

(viii) The proof of this identity is left to the reader as an exercise.

EXAMPLE 8. Using the formula $\operatorname{grad} r^n = nr^{n-2}\mathbf{r}$ (see Exercise 11, p. 81), prove that

$$\nabla^2 r^n = n(n+1)r^{n-2} \quad (r \neq 0 \text{ if } n \leqslant 2).$$

Solution. We have

$$\nabla^2 r^n = \operatorname{div}\operatorname{grad} r^n$$
$$= \operatorname{div}(nr^{n-2}\mathbf{r})$$
$$= nr^{n-2}\operatorname{div}\mathbf{r}+n\mathbf{r}.\operatorname{grad} r^{n-2},$$

using identity (4.49). Hence

$$\nabla^2 r^n = 3nr^{n-2}+n(n-2)r^{n-4}\mathbf{r}.\mathbf{r}$$
$$= n(n+1)r^{n-2}.$$

EXAMPLE 9. Show that the vector field $\mathbf{H}=\Phi\nabla\Psi$ is perpendicular to $\operatorname{curl}\mathbf{H}$ at all points where neither vector field vanishes.

Solution. Using identity (4.58),

$$\nabla\times(\Phi\nabla\Psi) = \Phi\nabla\times\nabla\Psi-\nabla\Psi\times\nabla\Phi$$
$$= \nabla\Phi\times\nabla\Psi, \quad \text{by (4.55).}$$

Thus,

$$\mathbf{H}.\operatorname{curl}\mathbf{H} = \Phi\nabla\Psi.(\nabla\Phi\times\nabla\Psi)$$
$$= 0,$$

showing that $\mathbf{H}$ and $\operatorname{curl}\mathbf{H}$ are perpendicular at points where $\mathbf{H}\neq\mathbf{0}$ and $\operatorname{curl}\mathbf{H}\neq\mathbf{0}$.

EXERCISES

27. Prove the identities (4.50) and (4.53).

28. Prove that $\operatorname{curl}(\mathbf{r}/r^2)=\mathbf{0}$. Also, show that $\operatorname{div}(\mathbf{r}/r^2)=1/r^2$.

29. Using the notation introduced at the beginning of this section, show that

$$\nabla^2 \equiv \mathbf{e}_i.\mathbf{e}_j\,\partial^2/\partial x_i\,\partial x_j.$$

Using expression (4.44) for curl, prove the identity

$$\operatorname{curl}\operatorname{curl}\mathbf{F} \equiv \operatorname{grad}(\operatorname{div}\mathbf{F})-\nabla^2\mathbf{F}.$$

30. Show that

$$\operatorname{curl}(\mathbf{r}\times\operatorname{curl}\mathbf{F})+(\mathbf{r}.\nabla)\operatorname{curl}\mathbf{F}+2\operatorname{curl}\mathbf{F} \equiv \mathbf{0}.$$

31. Prove that

$$\nabla^2(\Phi\Psi) \equiv \Phi\nabla^2\Psi+2\nabla\Phi.\nabla\Psi+\Psi\nabla^2\Phi.$$

32. If Φ satisfies Laplace's equation $\nabla^2\Phi=0$ and is such that $\mathbf{r}.\nabla\Phi=m\Phi$, where m is constant, prove that Φ/r^{2m+1} is also a solution of Laplace's equation. [*Hint.* Put $\Psi=1/r^{2m+1}$ in Exercise 31, and use the formula of Exercise 11, p. 81. See also Example 8, p. 97.]

4.10 Cylindrical and spherical polar coordinates

In the theory developed so far, we have worked entirely with rectangular cartesian coordinates. However, in practice other coordinate systems are often more convenient. Cylindrical polar coordinates and spherical polar coordinates are especially important.

Cylindrical polar coordinates. Let a point P have rectangular cartesian coordinates (x, y, z). Denote by R the perpendicular distance from P to the z-axis, and let ϕ be the angle between the zx-plane and the plane containing P and the axis Oz; we count ϕ positive in the sense indicated by the arrow in Fig. 37, and restrict its values to the range $-\pi < \phi \leqslant \pi$, or sometimes $0 \leqslant \phi < 2\pi$. Then R, ϕ, z are the *cylindrical polar coordinates* of P.

From the geometry of Fig. 37, the rectangular coordinates (x, y, z) of P are related to the cylindrical polar coordinates (R, ϕ, z) by the equations

$$x = R\cos\phi, \quad y = R\sin\phi, \quad z = z. \tag{4.62}$$

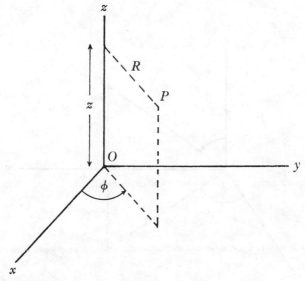

FIG. 37. Cylindrical polar coordinates (R, ϕ, z)

The equations can be solved for R, ϕ giving

$$\left.\begin{array}{l} R = (x^2 + y^2)^{\frac{1}{2}}, \\ \phi = \tan^{-1}(y/x). \end{array}\right\} \tag{4.63}$$

Spherical polar coordinates. As before, let P have rectangular cartesian coordinates (x, y, z). Let r denote the distance of P from the origin O, let θ be the angle that OP makes with the z-axis, and let ϕ be the angle between the zx-plane and the plane containing P and the z-axis. The angles θ, ϕ are measured positive in the senses shown by the arrows in Fig. 38, and their values are restricted to the ranges $0 \leqslant \theta \leqslant \pi$, $-\pi < \phi \leqslant \pi$ (or sometimes $0 \leqslant \phi < 2\pi$). Then r, θ, ϕ are the *spherical polar coordinates* of P.

From the geometry of Fig. 38, the rectangular cartesian coordinates

(x, y, z) of P are related to the spherical polar coordinates (r, θ, ϕ) by the equations

$$x = r\sin\theta\cos\phi, \quad y = r\sin\theta\sin\phi, \quad z = r\cos\theta. \qquad (4.64)$$

Solving for r, θ, ϕ in terms of x, y, z, we have

$$r = (x^2 + y^2 + z^2)^{\frac{1}{2}},$$

$$\theta = \cos^{-1}\frac{z}{(x^2 + y^2 + z^2)^{\frac{1}{2}}}, \qquad (4.65)$$

$$\phi = \tan^{-1}(y/x).$$

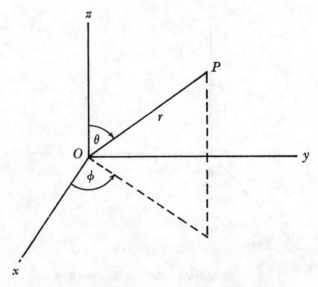

FIG. 38. Spherical polar coordinates (r, θ, ϕ)

Coordinate lines. If the coordinates y, z of the point $P(x, y, z)$ are fixed, and x is allowed to range over its permissible values, the locus of P is a straight line parallel to the x-axis: it is called an *x-coordinate line*. Similarly if x, z are fixed and y varies, the locus is a *y-coordinate line*; and if x, y are fixed and z varies, the locus is a *z-coordinate line* (Fig. 39).

In the case of cylindrical polar coordinates, the locus of $P(R, \phi, z)$ when ϕ, z are fixed and R varies is called an *R-coordinate line*; the locus when R, z are fixed and ϕ varies is a ϕ-coordinate line; and the locus when R, ϕ are fixed and z varies is a z-coordinate line. The R-coordinate lines are straight lines radiating from and normal to the z-axis; the ϕ-coordinate lines are circles centred on the z-axis and parallel to the xy-plane; and the

z-coordinate lines are straight lines parallel to the z-axis (Fig. 40a). The coordinate lines of a spherical polar coordinate system are defined similarly (Fig. 40b).

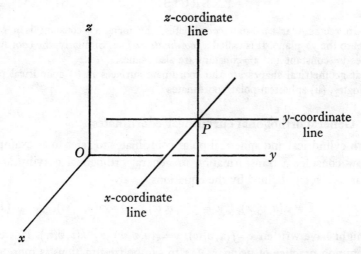

FIG. 39. Coordinate lines for rectangular cartesian coordinates

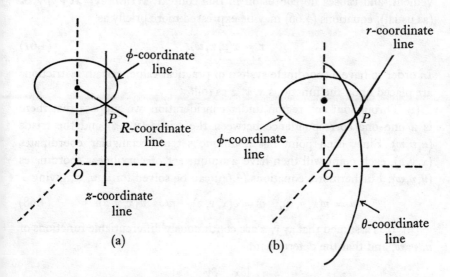

FIG. 40. Coordinate lines for (a) cylindrical and (b) spherical polar coordinates

In each of the above coordinate systems, the tangents at a point P to the three coordinate lines through P are mutually perpendicular. Such coordinate systems are said to be *orthogonal*. Because of this feature, it is easy to discuss them all within the framework of the theory of rectangular cartesian systems.

EXERCISE

33. In rectangular cartesian coordinates, the locus $x =$ constant is a plane parallel to the yz-plane. It is called a *coordinate surface*. Similarly, the coordinate surfaces $y =$ constant and $z =$ constant are also planes.

What geometrical shapes are the coordinate surfaces in (i) cylindrical polar coordinates, (ii) spherical polar coordinates?

4.11   General orthogonal curvilinear coordinates

With cylindrical and spherical polar coordinates to refer to as examples, we now consider a transformation to general orthogonal curvilinear coordinates (u, v, w), defined by the equations

$$x = x(u, v, w), \quad y = y(u, v, w), \quad z = z(u, v, w). \tag{4.66}$$

We might have written $x = f(u, v, w)$, $y = g(u, v, w)$, $z = h(u, v, w)$. However, the common practice of using x, y, z to symbolize functions is more convenient, and causes no confusion in this context. Writing $\mathbf{r} = x\mathbf{i} + y\mathbf{j} + z\mathbf{k}$ (as usual), equations (4.66) may be expressed more briefly as

$$\mathbf{r} = \mathbf{r}(u, v, w). \tag{4.67}$$

In order to have a coordinate system of practical value, certain restrictions are placed upon the functions x, y, z as follows.

(i) Throughout any region under consideration, we suppose that there is a one–one correspondence between the triads (x, y, z) and the triads (u, v, w). Since each point has a unique set of rectangular coordinates (x, y, z), each point will then have a unique set of curvilinear coordinates (u, v, w). Furthermore, equations (4.66) can be solved for u, v, w, giving

$$u = u(x, y, z), \quad v = v(x, y, z), \quad w = w(x, y, z). \tag{4.68}$$

(ii) It is assumed that x, y, z are continuously differentiable functions of u, v, w, and that the determinant

$$J = \begin{vmatrix} \dfrac{\partial x}{\partial u} & \dfrac{\partial y}{\partial u} & \dfrac{\partial z}{\partial u} \\[2mm] \dfrac{\partial x}{\partial v} & \dfrac{\partial y}{\partial v} & \dfrac{\partial z}{\partial v} \\[2mm] \dfrac{\partial x}{\partial w} & \dfrac{\partial y}{\partial w} & \dfrac{\partial z}{\partial w} \end{vmatrix} \tag{4.69}$$

does not vanish at any point. This determinant is called the *Jacobian* of the transformation.

In addition to these conditions, the tangents at a point P to the u-, v-, w-coordinate lines through this point are assumed to be mutually perpendicular, so that the coordinate system is *orthogonal*. The parametric equations of the coordinate lines are obtained by assigning constant values to pairs of u, v, w in equation (4.67). Thus, if u_0, v_0, w_0 are constants, the equations

$$\mathbf{r} = \mathbf{r}(u, v_0, w_0), \quad \mathbf{r} = \mathbf{r}(u_0, v, w_0), \quad \mathbf{r} = \mathbf{r}(u_0, v_0, w) \qquad (4.70)$$

represent u-, v-, w-coordinate lines, respectively.

Transformations encountered in practice are usually chosen to satisfy the above conditions everywhere except possibly at isolated points or on certain curves or surfaces. Thus, in the case of cylindrical polar coordinates, the one–one correspondence between the triads (x, y, z) and the triads (R, ϕ, z) fails on the z-axis; the values $R = 0$, $z = z_0$ correspond to the fixed point $x = y = 0$, $z = z_0$ on the z-axis regardless of the value of ϕ. However, such exceptions are easily recognized and will usually cause no serious difficulty. They will not be considered further.

Consider the coordinate transformation defined by equations (4.67), and let

$$h_1 = \left| \frac{\partial \mathbf{r}}{\partial u} \right|, \quad h_2 = \left| \frac{\partial \mathbf{r}}{\partial v} \right|, \quad h_3 = \left| \frac{\partial \mathbf{r}}{\partial w} \right|. \qquad (4.71)$$

The elements of the first, second, and third rows of the Jacobian (4.69) are the components of $\partial \mathbf{r}/\partial u$, $\partial \mathbf{r}/\partial v$, $\partial \mathbf{r}/\partial w$ respectively. As the Jacobian is assumed to be non-vanishing, at least one element of each row must be non-zero, and so it follows that h_1, h_2, h_3 do not vanish.

Let $\mathbf{e}_u$, $\mathbf{e}_v$, $\mathbf{e}_w$ respectively denote the unit tangents at a point P to the u-, v-, w-coordinate lines through P. Then (cf. § 3.4)

$$\mathbf{e}_u = \frac{1}{h_1} \frac{\partial \mathbf{r}}{\partial u}, \quad \mathbf{e}_v = \frac{1}{h_2} \frac{\partial \mathbf{r}}{\partial v}, \quad \mathbf{e}_w = \frac{1}{h_3} \frac{\partial \mathbf{r}}{\partial w}, \qquad (4.72)$$

where the right-hand sides are, of course, evaluated at P. These unit vectors will be mutually perpendicular, since the coordinate system is orthogonal. For convenience, the parameters u, v, w of the coordinate lines are always chosen so that $\mathbf{e}_u$, $\mathbf{e}_v$, $\mathbf{e}_w$ (in that order) form a right-handed triad. This can always be achieved, because if u is replaced by $-u$ in (4.67) the direction of $\partial \mathbf{r}/\partial u$ is then reversed.

The vectors $\mathbf{e}_u$, $\mathbf{e}_v$, $\mathbf{e}_w$ form an orthonormal triad, just as do the vectors $\mathbf{i}$, $\mathbf{j}$, $\mathbf{k}$. There is, however, one fundamental difference between the two

8

triads: whereas $\mathbf{i}, \mathbf{j}, \mathbf{k}$ point in fixed directions, *the directions of $\mathbf{e}_u, \mathbf{e}_v, \mathbf{e}_w$ will in general vary from point to point*, because the coordinate lines will be curved (Fig. 41).

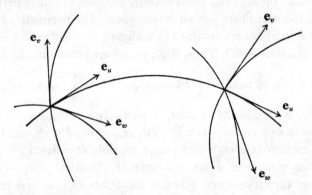

FIG. 41. Showing the unit vectors $\mathbf{e}_u, \mathbf{e}_v, \mathbf{e}_w$ at two different points on a
u-coordinate line

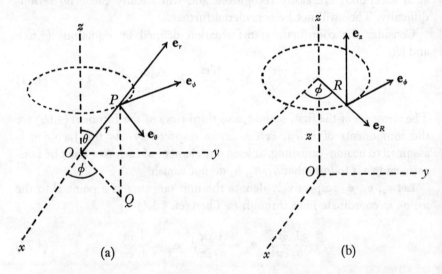

FIG. 42. (a) Unit vectors $\mathbf{e}_r, \mathbf{e}_\theta, \mathbf{e}_\phi$ for spherical polar coordinates
(r, θ, ϕ). $\mathbf{e}_\theta$ lies in the plane OPQ and $\mathbf{e}_\phi$ is perpendicular to OPQ.
(b) Unit vectors $\mathbf{e}_R, \mathbf{e}_\phi, \mathbf{e}_z$ for cylindrical polar coordinates (R, ϕ, z).
$\mathbf{e}_R$ and $\mathbf{e}_\phi$ are parallel to the xy-plane

EXAMPLE 10 (*Spherical polar coordinates*). In the case of spherical polar coordinates defined by equations (4.64), we have

$$\mathbf{r} = r(\sin\theta\cos\phi, \sin\theta\sin\phi, \cos\theta). \tag{4.73}$$

Therefore,

$$\frac{\partial \mathbf{r}}{\partial r} = (\sin\theta\cos\phi, \sin\theta\sin\phi, \cos\theta),$$

$$\frac{\partial \mathbf{r}}{\partial \theta} = r(\cos\theta\cos\phi, \cos\theta\sin\phi, -\sin\theta),$$

and

$$\frac{\partial \mathbf{r}}{\partial \phi} = r(-\sin\theta\sin\phi, \sin\theta\cos\phi, 0).$$

It follows that

$$h_1^2 = \sin^2\theta\cos^2\phi + \sin^2\theta\sin^2\phi + \cos^2\theta$$
$$= \sin^2\theta(\cos^2\phi + \sin^2\phi) + \cos^2\theta$$
$$= 1.$$

Similarly,

$$h_2^2 = r^2 \quad \text{and} \quad h_3^2 = r^2\sin^2\theta.$$

Since $0 \leqslant \theta \leqslant \pi$, we have $\sin\theta \geqslant 0$. Thus

$$h_1 = 1, \quad h_2 = r, \quad h_3 = r\sin\theta. \tag{4.74}$$

From the definition (4.71), it follows also that

$$\begin{aligned}
\mathbf{e}_r &= (\sin\theta\cos\phi, \sin\theta\sin\phi, \cos\theta), \\
\mathbf{e}_\theta &= (\cos\theta\cos\phi, \cos\theta\sin\phi, -\sin\theta), \\
\mathbf{e}_\phi &= (-\sin\phi, \cos\phi, 0).
\end{aligned} \tag{4.75}$$

The directions of these unit vectors are shown in Fig. 42(a).

EXAMPLE 11 (*Cylindrical polar coordinates*). In the case of cylindrical polar coordinates (R, ϕ, z), defined by equations (4.62), it is found that

$$h_1 = 1, \quad h_2 = R, \quad h_3 = 1. \tag{4.76}$$

Also,

$$\begin{aligned}
\mathbf{e}_R &= (\cos\phi, \sin\phi, 0), \\
\mathbf{e}_\phi &= (-\sin\phi, \cos\phi, 0), \\
\mathbf{e}_z &= (0, 0, 1).
\end{aligned} \tag{4.77}$$

It is left as an exercise for the reader to establish these results. Fig. 42(b) shows the directions of the unit vectors $\mathbf{e}_R$, $\mathbf{e}_\phi$, $\mathbf{e}_z$.

The formulae for h_1, h_2, h_3, can sometimes be obtained by the following geometrical argument. Suppose that u is given an infinitesimal positive increment du, and that v, w are held constant. Then, from (4.67), the corresponding increment in $\mathbf{r}$ is

$$d\mathbf{r} = \frac{\partial \mathbf{r}}{\partial u} du,$$

and so

$$|d\mathbf{r}| = \left|\frac{\partial \mathbf{r}}{\partial u}\right| du = h_1 du.$$

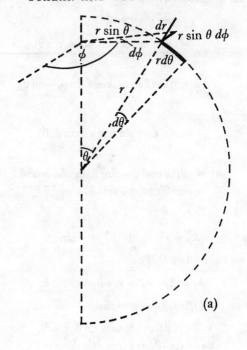

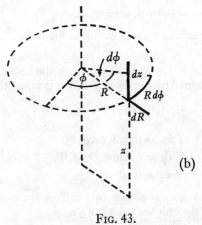

FIG. 43.

(a) In spherical polar coordinates (r, θ, ϕ)

$$d\mathbf{r} = (dr, r d\theta, r \sin \theta \, d\phi)$$

Thus $h_1 = 1$, $h_2 = r$, $h_3 = r \sin \theta$

(b) In cylindrical polar coordinates (R, ϕ, z)

$$d\mathbf{r} = (dR, R d\phi, dz)$$

Thus $h_1 = 1$, $h_2 = R$, $h_3 = 1$

Since $\mathbf{r}$ is the position vector of a point P relative to the origin, $|d\mathbf{r}| = h_1 \, du$ is the displacement of P corresponding to the increment du in u. Similarly, $h_2 \, dv$ and $h_3 \, dw$ are the displacements corresponding to increments dv in v and dw in w respectively. Thus h_1, h_2, h_3 may be found by giving small increments to u, v, w and considering the geometrical effect. Fig. 43 indicates how this method may be applied in the case of spherical polar coordinates and cylindrical polar coordinates.

EXERCISES

34. Prove the results stated in equations (4.76) and (4.77).

35. *Toroidal curvilinear coordinates* R, θ, ϕ are defined in terms of rectangular cartesian coordinates x, y, z by

$$x = (a - R\cos\theta)\cos\phi, \quad y = (a - R\cos\theta)\sin\phi, \quad z = R\sin\theta,$$

where a is a constant and $R < a$. Show that this is an orthogonal coordinate system and that, in the usual notation,

$$h_1 = 1, \quad h_2 = R, \quad h_3 = a - R\cos\theta.$$

36. *Elliptical coordinates* ξ, η, z are such that the position vector is given by

$$\mathbf{r} = (\cosh\xi\cos\eta, \sinh\xi\sin\eta, z),$$

where $0 \leqslant \xi < \infty$, $-\pi < \eta \leqslant \pi$, and $-\infty < z < \infty$. Show that this curvilinear system of coordinates is orthogonal, and find h_1, h_2, h_3. What are the geometrical shapes of the ξ- and η-coordinate lines?

37. *Parabolic coordinates* u, v, w are defined so that the position vector relative to rectangular cartesian axes is

$$\mathbf{r} = (\tfrac{1}{2}(u^2 - v^2), uv, -w),$$

where $-\infty < u < \infty$, $0 \leqslant v < \infty$, $-\infty < w < \infty$. Show that, for any fixed value of w, the u- and v-coordinate curves are confocal parabolas. Verify that the system is orthogonal, and show that the basic unit vectors at the point $u = v = w = 1$ are $(1, 1, 0)/\sqrt{2}, (-1, 1, 0)/\sqrt{2}$, and $(0, 0, -1)$. Show that cylindrical polar coordinates R, ϕ, z may be chosen so that

$$u = (2R)^{\frac{1}{2}}\cos\tfrac{1}{2}\phi, \quad v = (2R)^{\frac{1}{2}}\sin\tfrac{1}{2}\phi, \quad w = -z.$$

38. In spherical polar coordinates, show that

$$\partial\mathbf{e}_r/\partial\theta = \mathbf{e}_\theta, \quad \partial\mathbf{e}_\theta/\partial\theta = -\mathbf{e}_r, \quad \partial\mathbf{e}_\phi/\partial\theta = \mathbf{0},$$
$$\partial\mathbf{e}_r/\partial\phi = \sin\theta\,\mathbf{e}_\phi, \quad \partial\mathbf{e}_\theta/\partial\phi = \cos\theta\,\mathbf{e}_\phi,$$
$$\partial\mathbf{e}_\phi/\partial\phi = -\sin\theta\,\mathbf{e}_r - \cos\theta\,\mathbf{e}_\theta.$$

4.12   Vector components in orthogonal curvilinear coordinates

Suppose that a vector field has the value $\mathbf{F}$ at the point P with curvilinear coordinates (u, v, w). Since the unit vectors $\mathbf{e}_u$, $\mathbf{e}_v$, $\mathbf{e}_w$ are not coplanar, it follows from the worked example 5 on p. 44 that $\mathbf{F}$ can be expressed in the form

$$\mathbf{F} = F_u \mathbf{e}_u + F_v \mathbf{e}_v + F_w \mathbf{e}_w. \tag{4.78}$$

In other words, the vectors $\mathbf{e}_u$, $\mathbf{e}_v$, $\mathbf{e}_w$ associated with P can be taken as an orthonormal basis for the representation of $\mathbf{F}$. We call F_u, F_v, F_w the *components of* $\mathbf{F}$ *along the coordinate lines*.

When rectangular cartesian coordinates were used, we wrote

$$\mathbf{F} = (F_1, F_2, F_3),$$

and it may seem natural now to use the notation

$$\mathbf{F} = (F_u, F_v, F_w). \tag{4.79}$$

However, in this case the variation of $\mathbf{F}$ with position is not represented completely by the variation of F_u, F_v, F_w, because the base vectors $\mathbf{e}_u$, $\mathbf{e}_v$, $\mathbf{e}_w$ are themselves functions of u, v, w in general. Thus,

$$\frac{\partial \mathbf{F}}{\partial u} = \frac{\partial F_u}{\partial u}\mathbf{e}_u + \frac{\partial F_v}{\partial u}\mathbf{e}_v + \frac{\partial F_w}{\partial u}\mathbf{e}_w + F_u \frac{\partial \mathbf{e}_u}{\partial u} + F_v \frac{\partial \mathbf{e}_v}{\partial u} + F_w \frac{\partial \mathbf{e}_w}{\partial u},$$

whence in general

$$\frac{\partial \mathbf{F}}{\partial u} \neq \left(\frac{\partial F_u}{\partial u}, \frac{\partial F_v}{\partial u}, \frac{\partial F_w}{\partial u} \right).$$

In order to avoid possible misunderstanding, the notation (4.78) will always be used in preference to (4.79) when the coordinate system is curvilinear.

The procedure for expressing a vector field

$$\mathbf{F}(x, y, z) = F_1(x, y, z)\mathbf{i} + F_2(x, y, z)\mathbf{j} + F_3(x, y, z)\mathbf{k} \tag{4.80}$$

in the form

$$\mathbf{F}(u, v, w) = F_u(u, v, w)\mathbf{e}_u + F_v(u, v, w)\mathbf{e}_v + F_w(u, v, w)\mathbf{e}_w \tag{4.81}$$

is straightforward in principle. First, F_1, F_2, F_3 can be expressed as functions of u, v, w by substitution from equations (4.66). Also, since the triad $\mathbf{e}_u$, $\mathbf{e}_v$, $\mathbf{e}_w$ is orthonormal,

$$F_u = \mathbf{F} . \mathbf{e}_u = F_1 \mathbf{i} . \mathbf{e}_u + F_2 \mathbf{j} . \mathbf{e}_u + F_3 \mathbf{k} . \mathbf{e}_u.$$

But from the defining equations (4.71) and (4.72),

$$\mathbf{e}_u = \frac{1}{h_1}\frac{\partial \mathbf{r}}{\partial u} = \frac{1}{h_1}\left(\frac{\partial x}{\partial u}\mathbf{i} + \frac{\partial y}{\partial u}\mathbf{j} + \frac{\partial z}{\partial u}\mathbf{k} \right),$$

where

$$h_1 = \left\{ \left(\frac{\partial x}{\partial u}\right)^2 + \left(\frac{\partial y}{\partial u}\right)^2 + \left(\frac{\partial z}{\partial u}\right)^2 \right\}^{\frac{1}{2}}.$$

Hence,

$$F_u = \frac{1}{h_1} \left(F_1 \frac{\partial x}{\partial u} + F_2 \frac{\partial y}{\partial u} + F_3 \frac{\partial z}{\partial u} \right).$$

Similarly, F_v and F_w can be found in terms of u, v, w, and $\mathbf{F}$ is thus represented in the form (4.81).

EXAMPLE 12. Express the vector field

$$\mathbf{F} = z\mathbf{i} = (z, 0, 0)$$

in the form

$$\mathbf{F} = F_r \mathbf{e}_r + F_\theta \mathbf{e}_\theta + F_\phi \mathbf{e}_\phi,$$

where r, θ, ϕ are spherical polar coordinates.

Solution. The unit vectors $\mathbf{e}_r$, $\mathbf{e}_\theta$, $\mathbf{e}_\phi$ have already been obtained, and are given by equations (4.75). Thus, using those equations, and also the relation $z = r\cos\theta$,

$$F_r = \mathbf{F} \cdot \mathbf{e}_r = z \sin\theta \cos\phi = r \cos\theta \sin\theta \cos\phi,$$
$$F_\theta = \mathbf{F} \cdot \mathbf{e}_\theta = z \cos\theta \cos\phi = r \cos^2\theta \cos\phi,$$
$$F_\phi = \mathbf{F} \cdot \mathbf{e}_\phi = -z \sin\phi = -r \cos\theta \sin\phi.$$

Hence

$$\mathbf{F} = r \cos\theta \sin\theta \cos\phi \, \mathbf{e}_r + r \cos^2\theta \cos\phi \, \mathbf{e}_\theta - r \cos\theta \sin\phi \, \mathbf{e}_\phi.$$

EXERCISES

39. Express the vector field $\mathbf{F} = (-y, x, 0)$ in cylindrical polar and in spherical polar component form.

40. Express the position vector in cylindrical polar coordinate form.

41. In spherical polar coordinates,

$$\mathbf{F} = r\mathbf{e}_\theta + r\mathbf{e}_\phi.$$

Find the spherical polar components of $\partial \mathbf{F}/\partial r$, $\partial \mathbf{F}/\partial \theta$ and $\partial \mathbf{F}/\partial \phi$.

4.13 Expressions for $\operatorname{grad}\Omega$, $\operatorname{div}\mathbf{F}$, $\operatorname{curl}\mathbf{F}$, and ∇^2 in orthogonal curvilinear coordinates

Using the notation of §§ 4.11 and 4.12, the following formulae will now be established:

(i)
$$\operatorname{grad}\Omega = \frac{\mathbf{e}_u}{h_1}\frac{\partial\Omega}{\partial u} + \frac{\mathbf{e}_v}{h_2}\frac{\partial\Omega}{\partial v} + \frac{\mathbf{e}_w}{h_3}\frac{\partial\Omega}{\partial w} ; \qquad (4.82)$$

or alternatively, in operator notation,

$$\boldsymbol{\nabla} = \frac{\mathbf{e}_u}{h_1}\frac{\partial}{\partial u} + \frac{\mathbf{e}_v}{h_2}\frac{\partial}{\partial v} + \frac{\mathbf{e}_w}{h_3}\frac{\partial}{\partial w} . \qquad (4.83)$$

(ii) $\mathrm{div}\,\mathbf{F} = \dfrac{1}{h_1 h_2 h_3}\left\{\dfrac{\partial}{\partial u}(h_2 h_3 F_u)+\dfrac{\partial}{\partial v}(h_3 h_1 F_v)+\dfrac{\partial}{\partial w}(h_1 h_2 F_w)\right\}.$ (4.84)

(iii) $\mathrm{curl}\,\mathbf{F} = \dfrac{1}{h_1 h_2 h_3}\begin{vmatrix} h_1\,\mathbf{e}_u & h_2\,\mathbf{e}_v & h_3\,\mathbf{e}_w \\[4pt] \dfrac{\partial}{\partial u} & \dfrac{\partial}{\partial v} & \dfrac{\partial}{\partial w} \\[8pt] h_1 F_u & h_2 F_v & h_3 F_w \end{vmatrix}.$ (4.85)

(iv) $\nabla^2 = \dfrac{1}{h_1 h_2 h_3}\left\{\dfrac{\partial}{\partial u}\left(\dfrac{h_2 h_3}{h_1}\dfrac{\partial}{\partial u}\right)+\dfrac{\partial}{\partial v}\left(\dfrac{h_3 h_1}{h_2}\dfrac{\partial}{\partial v}\right)+\dfrac{\partial}{\partial w}\left(\dfrac{h_1 h_2}{h_3}\dfrac{\partial}{\partial w}\right)\right\}.$ (4.86)

N.B. The formulae in the particular cases of cylindrical and spherical polar coordinates are listed in Appendix 3 for easy reference.

Proofs.
(i) We have

$$\mathbf{e}_u = \frac{1}{h_1}\frac{\partial \mathbf{r}}{\partial u} = \frac{1}{h_1}\left(\frac{\partial x}{\partial u},\,\frac{\partial y}{\partial u},\,\frac{\partial z}{\partial u}\right)$$

and

$$\mathrm{grad}\,\Omega = \left(\frac{\partial \Omega}{\partial x},\,\frac{\partial \Omega}{\partial y},\,\frac{\partial \Omega}{\partial z}\right).$$

Hence the component of $\mathrm{grad}\,\Omega$ in the direction of $\mathbf{e}_u$ is

$$\mathbf{e}_u\cdot\mathrm{grad}\,\Omega = \frac{1}{h_1}\left(\frac{\partial \Omega}{\partial x}\frac{\partial x}{\partial u}+\frac{\partial \Omega}{\partial y}\frac{\partial y}{\partial u}+\frac{\partial \Omega}{\partial z}\frac{\partial z}{\partial u}\right)$$

$$= \frac{1}{h_1}\frac{\partial \Omega}{\partial u}\quad \text{(by the chain rule (4.3)).}\qquad (4.87)$$

The components of $\mathrm{grad}\,\Omega$ in the directions $\mathbf{e}_v$, $\mathbf{e}_w$ follow similarly, and hence

$$\mathrm{grad}\,\Omega = \frac{\mathbf{e}_u}{h_1}\frac{\partial \Omega}{\partial u}+\frac{\mathbf{e}_v}{h_2}\frac{\partial \Omega}{\partial v}+\frac{\mathbf{e}_w}{h_3}\frac{\partial \Omega}{\partial w}.$$

(ii) It is convenient to use operator notation in this proof.
We note first that taking Ω to be u, v, w in turn in (4.82), gives

$$\mathbf{e}_u = h_1\nabla u,\quad \mathbf{e}_v = h_2\nabla v,\quad \mathbf{e}_w = h_3\nabla w.\qquad (4.88)$$

Thus, since $\mathbf{e}_u$, $\mathbf{e}_v$, $\mathbf{e}_w$ are a right-handed orthonormal triad,

$$\mathbf{e}_u = \mathbf{e}_v\times\mathbf{e}_w = h_2 h_3\,\nabla v\times\nabla w.$$

Now

$$\nabla\cdot(F_u\mathbf{e}_u) = \nabla\cdot(h_2 h_3 F_u\,\nabla v\times\nabla w)$$
$$= h_2 h_3 F_u\nabla\cdot(\nabla v\times\nabla w)+(\nabla v\times\nabla w)\cdot\nabla(h_2 h_3 F_u),$$

using identity (4.57). But using (4.60) followed by (4.55),

$$\mathbf{\nabla}.(\mathbf{\nabla}v \times \mathbf{\nabla}w) = \mathbf{\nabla}w.(\mathbf{\nabla} \times \mathbf{\nabla}v) - \mathbf{\nabla}v.(\mathbf{\nabla} \times \mathbf{\nabla}w)$$
$$= 0.$$

Hence

$$\mathbf{\nabla}.(F_u\mathbf{e}_u) = (\mathbf{\nabla}v \times \mathbf{\nabla}w).\mathbf{\nabla}(h_2h_3F_u)$$

$$= \frac{\mathbf{e}_u}{h_2h_3}.\mathbf{\nabla}(h_2h_3F_u)$$

$$= \frac{1}{h_1h_2h_3}\frac{\partial}{\partial u}(h_2h_3F_u), \quad \text{using (4.87).}$$

Similar results can be obtained for $\mathbf{\nabla}.(F_v\mathbf{e}_v)$ and $\mathbf{\nabla}.(F_w\mathbf{e}_w)$. Thus, since

$$\mathbf{\nabla}.(F_u\mathbf{e}_u)+\mathbf{\nabla}.(F_v\mathbf{e}_v)+\mathbf{\nabla}.(F_w\mathbf{e}_w) = \mathbf{\nabla}.(F_u\mathbf{e}_u+F_v\mathbf{e}_v+F_w\mathbf{e}_w)$$
$$= \text{div}\,\mathbf{F},$$

formula (4.84) follows.

(iii) We have

$$\mathbf{\nabla} \times (F_u\mathbf{e}_u) = \mathbf{\nabla} \times (h_1F_u\mathbf{\nabla}u) \quad \text{(using (4.88))}$$

$$= h_1F_u\mathbf{\nabla} \times (\mathbf{\nabla}u) - \mathbf{\nabla}u \times \mathbf{\nabla}(h_1F_u) \quad \text{(using identity (4.58))}$$

$$= \mathbf{\nabla}(h_1F_u) \times \mathbf{\nabla}u \quad \text{(using identity (4.55))}$$

$$= \left[\frac{\mathbf{e}_u}{h_1}\frac{\partial}{\partial u}(h_1F_u)+\frac{\mathbf{e}_v}{h_2}\frac{\partial}{\partial v}(h_1F_u)+\frac{\mathbf{e}_w}{h_3}\frac{\partial}{\partial w}(h_1F_u)\right] \times \frac{\mathbf{e}_u}{h_1}$$

$$= \frac{\mathbf{e}_v}{h_1h_3}\frac{\partial}{\partial w}(h_1F_u)-\frac{\mathbf{e}_w}{h_1h_2}\frac{\partial}{\partial v}(h_1F_u)$$

$$= \frac{1}{h_1h_2h_3}\begin{vmatrix} h_1\mathbf{e}_u & h_2\mathbf{e}_v & h_3\mathbf{e}_w \\ \dfrac{\partial}{\partial u} & \dfrac{\partial}{\partial v} & \dfrac{\partial}{\partial w} \\ h_1F_u & 0 & 0 \end{vmatrix}.$$

Similarly,

$$\mathbf{\nabla} \times (F_v\mathbf{e}_v) = \frac{1}{h_1h_2h_3}\begin{vmatrix} h_1\mathbf{e}_u & h_2\mathbf{e}_v & h_3\mathbf{e}_w \\ \dfrac{\partial}{\partial u} & \dfrac{\partial}{\partial v} & \dfrac{\partial}{\partial w} \\ 0 & h_2F_v & 0 \end{vmatrix}$$

and

$$\mathbf{\nabla} \times (F_w\mathbf{e}_w) = \frac{1}{h_1h_2h_3}\begin{vmatrix} h_1\mathbf{e}_u & h_2\mathbf{e}_v & h_3\mathbf{e}_w \\ \dfrac{\partial}{\partial u} & \dfrac{\partial}{\partial v} & \dfrac{\partial}{\partial w} \\ 0 & 0 & h_3F_w \end{vmatrix}.$$

Addition of the three results gives formula (4.85).

(iv) We have

$$\nabla^2 = \nabla . \nabla = \nabla . \left[\frac{\mathbf{e}_u}{h_1} \frac{\partial}{\partial u} + \frac{\mathbf{e}_v}{h_2} \frac{\partial}{\partial v} + \frac{\mathbf{e}_w}{h_3} \frac{\partial}{\partial w} \right].$$

Hence, applying (4.84) with

$$F_u, F_v, F_w \quad \text{replaced by} \quad \frac{1}{h_1} \frac{\partial}{\partial u}, \frac{1}{h_2} \frac{\partial}{\partial v}, \frac{1}{h_3} \frac{\partial}{\partial w}$$

respectively, formula (4.86) follows.

EXAMPLE 13. Find the curl and divergence of the vector field

$$\mathbf{H} = r^2 \cos \theta \, \mathbf{e}_r + (\mathbf{e}_\theta \sin \theta + \mathbf{e}_\phi)/r \sin \theta$$

in spherical polar coordinates r, θ, ϕ.

Solution. In spherical polar coordinates

$$h_1 = 1, \quad h_2 = r, \quad h_3 = r \sin \theta.$$

Thus

$$\operatorname{curl} \mathbf{H} = \frac{1}{r^2 \sin \theta} \begin{vmatrix} \mathbf{e}_r & r \mathbf{e}_\theta & r \sin \theta \, \mathbf{e}_\phi \\ \partial/\partial r & \partial/\partial \theta & \partial/\partial \phi \\ r^2 \cos \theta & 1 & 1 \end{vmatrix}$$

$$= r \sin \theta \, \mathbf{e}_\phi.$$

Also,

$$\operatorname{div} \mathbf{H} = \frac{1}{r^2} \frac{\partial}{\partial r} (r^2 H_r) + \frac{1}{r \sin \theta} \frac{\partial}{\partial \theta} (H_\theta \sin \theta) + \frac{1}{r \sin \theta} \frac{\partial H_\phi}{\partial \phi}$$

$$= 4r \cos \theta + r^{-2} \cot \theta.$$

EXAMPLE 14. If $\psi(R, z)$ is a function of R and z only, in cylindrical polar coordinates R, ϕ, z, show that the vector field

$$\mathbf{H} = -\frac{1}{R} \frac{\partial \psi}{\partial z} \mathbf{e}_R + \frac{1}{R} \frac{\partial \psi}{\partial R} \mathbf{e}_z$$

satisfies the equations

$$\operatorname{div} \mathbf{H} = 0$$

and

$$\operatorname{curl} \mathbf{H} = \left(\frac{1}{R^2} \frac{\partial \psi}{\partial R} - \frac{1}{R} \frac{\partial^2 \psi}{\partial R^2} - \frac{1}{R} \frac{\partial^2 \psi}{\partial z^2} \right) \mathbf{e}_\phi.$$

Solution. In cylindrical polar coordinates

$$h_1 = 1, \quad h_2 = R, \quad h_3 = 1.$$

Thus

$$\operatorname{div}\mathbf{H} = \frac{1}{R}\frac{\partial}{\partial R}(RH_R) + \frac{1}{R}\frac{\partial H_\phi}{\partial \phi} + \frac{\partial H_z}{\partial z}$$

$$= -\frac{1}{R}\frac{\partial^2 \psi}{\partial R\, \partial z} + \frac{1}{R}\frac{\partial^2 \psi}{\partial z\, \partial R}$$

$$= 0,$$

provided that the order of differentiation is immaterial. Further,

$$\operatorname{curl}\mathbf{H} = \frac{1}{R}\begin{vmatrix} \mathbf{e}_R & R\,\mathbf{e}_\phi & \mathbf{e}_z \\ \partial/\partial R & \partial/\partial\phi & \partial/\partial z \\ -\dfrac{1}{R}\dfrac{\partial\psi}{\partial z} & 0 & \dfrac{1}{R}\dfrac{\partial\psi}{\partial R} \end{vmatrix}$$

$$= -\left[\frac{\partial}{\partial R}\left(\frac{1}{R}\frac{\partial\psi}{\partial R}\right) + \frac{1}{R}\frac{\partial^2 \psi}{\partial z^2}\right]\mathbf{e}_\phi,$$

giving the required result.

EXAMPLE 15. A vector field $\mathbf{F}$ is such that in terms of cylindrical polar coordinates, R, ϕ, z,

$$\mathbf{F} = F_R\mathbf{e}_R + F_\phi\mathbf{e}_\phi,$$

where F_R, F_ϕ are independent of z. Express $(\mathbf{F}.\nabla)\mathbf{F}$, $\frac{1}{2}\operatorname{grad}F^2$ and $\mathbf{F}\times\operatorname{curl}\mathbf{F}$ in terms of cylindrical polar components and verify that the identity

$$(\mathbf{F}.\nabla)\mathbf{F} = \tfrac{1}{2}\operatorname{grad} F^2 - \mathbf{F}\times\operatorname{curl}\mathbf{F} \qquad (4.89)$$

is satisfied.

Solution. In the case of cylindrical polar coordinates, $h_1 = 1$, $h_2 = R$, $h_3 = 1$ (equations (4.76)), and so

$$\nabla = \mathbf{e}_R\frac{\partial}{\partial R} + \mathbf{e}_\phi\frac{\partial}{R\,\partial\phi} + \mathbf{e}_z\frac{\partial}{\partial z}.$$

Hence

$$(\mathbf{F}.\nabla)\mathbf{F} = \left(F_R\frac{\partial}{\partial R} + F_\phi\frac{\partial}{R\partial\phi}\right)(F_R\mathbf{e}_R + F_\phi\mathbf{e}_\phi)$$

$$= F_R\frac{\partial F_R}{\partial R}\mathbf{e}_R + F_R^2\frac{\partial\mathbf{e}_R}{\partial R} + F_R\frac{\partial F_\phi}{\partial R}\mathbf{e}_\phi + F_R F_\phi\frac{\partial\mathbf{e}_\phi}{\partial R}$$

$$+ F_\phi\frac{\partial F_R}{R\,\partial\phi}\mathbf{e}_R + F_\phi F_R\frac{\partial\mathbf{e}_R}{R\,\partial\phi} + F_\phi\frac{\partial F_\phi}{R\,\partial\phi}\mathbf{e}_\phi + F_\phi^2\frac{\partial\mathbf{e}_\phi}{R\,\partial\phi}.$$

From equations (4.77),

$$\mathbf{e}_R = (\cos\phi, \sin\phi, 0), \quad \mathbf{e}_\phi = (-\sin\phi, \cos\phi, 0).$$

Hence

$$\frac{\partial \mathbf{e}_R}{\partial \phi} = (-\sin\phi, \cos\phi, 0) = \mathbf{e}_\phi,$$

$$\frac{\partial \mathbf{e}_\phi}{\partial \phi} = (-\cos\phi, -\sin\phi, 0) = -\mathbf{e}_R,$$

and the derivatives of $\mathbf{e}_R$ and $\mathbf{e}_\phi$ with respect to R are zero. Thus

$$(\mathbf{F}.\nabla)\mathbf{F} = \left(F_R\frac{\partial F_R}{\partial R} + F_\phi\frac{\partial F_R}{R\partial\phi} - \frac{F_\phi^2}{R}\right)\mathbf{e}_R$$

$$+ \left(F_R\frac{\partial F_\phi}{\partial R} + \frac{F_R F_\phi}{R} + F_\phi\frac{\partial F_\phi}{R\partial\phi}\right)\mathbf{e}_\phi. \qquad (4.90)$$

Also

$$\tfrac{1}{2}\,\text{grad}\,F^2 = \tfrac{1}{2}\left(\mathbf{e}_R\frac{\partial}{\partial R} + \mathbf{e}_\phi\frac{\partial}{R\partial\phi} + \mathbf{e}_z\frac{\partial}{\partial z}\right)(F_R^2 + F_\phi^2)$$

$$= \left(F_R\frac{\partial F_R}{\partial R} + F_\phi\frac{\partial F_\phi}{\partial R}\right)\mathbf{e}_R + \left(F_R\frac{\partial F_R}{R\partial\phi} + F_\phi\frac{\partial F_\phi}{R\partial\phi}\right)\mathbf{e}_\phi, \qquad (4.91)$$

since F_R, F_ϕ are independent of z.

Using formula (4.85),

$$\mathbf{F}\times\text{curl}\,\mathbf{F} = \mathbf{F}\times\frac{1}{R}\begin{vmatrix} \mathbf{e}_R & R\mathbf{e}_\phi & \mathbf{e}_z \\ \dfrac{\partial}{\partial R} & \dfrac{\partial}{\partial\phi} & \dfrac{\partial}{\partial z} \\ F_R & RF_\phi & 0 \end{vmatrix}$$

$$= \mathbf{F}\times\frac{1}{R}\left[\frac{\partial}{\partial R}(RF_\phi) - \frac{\partial F_R}{\partial\phi}\right]\mathbf{e}_z$$

$$= (F_R\mathbf{e}_R + F_\phi\mathbf{e}_\phi)\times\left(\frac{\partial F_\phi}{\partial R} + \frac{F_\phi}{R} - \frac{\partial F_R}{R\partial\phi}\right)\mathbf{e}_z$$

$$= \left(F_\phi\frac{\partial F_\phi}{\partial R} + \frac{F_\phi^2}{R} - F_\phi\frac{\partial F_R}{R\partial\phi}\right)\mathbf{e}_R$$

$$+ \left(F_R\frac{\partial F_R}{R\partial\phi} - \frac{F_R F_\phi}{R} - F_R\frac{\partial F_\phi}{\partial R}\right)\mathbf{e}_\phi. \qquad (4.92)$$

By substituting (4.90), (4.91) and (4.92) into (4.89), it is seen at once that the identity is satisfied. We note that (4.89) is a particular case of identity (4.51), viz. the case when $\mathbf{F} \equiv \mathbf{G}$.

EXERCISES

42. When u, v, w are identical to the rectangular cartesian coordinates x, y, z, show that $h_1 = h_2 = h_3 = 1$. Verify that, in this case, expressions (4.82)–(4.86) reduce to the original expressions whereby $\text{grad}\,\Omega$, $\text{div}\,\mathbf{F}$, $\text{curl}\,\mathbf{F}$, and ∇^2 were defined.

43. Find $\text{grad}(R^2 z \sin\phi\cos\phi)$ at the point $R=1$, $\phi=\tfrac{1}{4}\pi$, $z=2$.

44. The scalar field $\Omega(r,\theta,\phi)$ is such that

$$\mathbf{r}.\text{grad}\,\Omega = n\Omega,$$

where n is a constant. Show that Ω is of the form

$$\Omega = r^n f(\theta,\phi).$$

45. Find the divergence of the vector field $R\cos\phi\,\mathbf{e}_R + R\sin\phi\,\mathbf{e}_\phi$ in cylindrical polar coordinates R, ϕ, z.

46. In spherical polar coordinates, find the components of $\text{grad}(\text{div}\,\mathbf{e}_\theta)$.

47. Working entirely in cylindrical polar coordinates, verify that the vector field

$$\mathbf{F} = (R\sin\phi\cos\phi + z\cos\phi)\,\mathbf{e}_R + (R\cos^2\phi - z\sin\phi)\,\mathbf{e}_\phi + R\sin\phi\,\mathbf{e}_z$$

is such that $\text{curl}\,\text{curl}\,\mathbf{F} \equiv \mathbf{0}$.

48. In spherical polar coordinates, $\mathbf{F}=F_r\mathbf{e}_r+F_\theta\mathbf{e}_\theta$, where F_r, F_θ are independent of ϕ. Prove that $\text{curl}\,\text{curl}\,\mathbf{F}$ is of a similar form to $\mathbf{F}$.

49. Using spherical polar coordinates, show that if n is constant

$$\nabla^2 r^n = n(n+1)r^{n-2} \quad (r \neq 0 \text{ if } n \leqslant 2).$$

50. Show that, in cylindrical polar coordinates, Laplace's equation (4.37), p.93 is

$$\frac{\partial^2\Omega}{\partial R^2} + \frac{1}{R}\frac{\partial\Omega}{\partial R} + \frac{1}{R^2}\frac{\partial^2\Omega}{\partial\phi^2} + \frac{\partial^2\Omega}{\partial z^2} = 0.$$

51. If $\mathbf{F}=R\cos\phi\,\mathbf{e}_R+\sin\phi\,\mathbf{e}_\phi$, evaluate $(\mathbf{F}.\nabla)\mathbf{F}$.

52. Using cylindrical polar coordinates throughout, verify the identity

$$\text{curl}\,\text{curl}\,\mathbf{F} = \text{grad}(\text{div}\,\mathbf{F}) - \nabla^2\mathbf{F}$$

for the vector field $\mathbf{F}=\mathbf{e}_\phi$.

4.14 Vector analysis in n-dimensional space

The reader is probably aware of the possibility of generalising certain concepts associated with three-dimensional space to spaces of higher dimensions. Such generalisations are not, by any means, valueless; for example, in relativity theory, distance and time are found to be inextricably linked and it is necessary to work within the framework of a four-dimensional coordinate system.

This is a convenient point in our exposition of vector analysis to explain how some of the ideas with which we have been concerned so far may be extended to spaces of higher dimensions. The digression will be brief and must necessarily, therefore, be somewhat superficial.

In an n-dimensional space, the *points* are defined as ordered n-tuples of real numbers, of the form $(x_1, x_2, ..., x_n)$. If the distance between two points $P(x_1, x_2, ..., x_n)$ and $Q(y_1, y_2, ..., y_n)$ is defined as

$$d = [(x_1 - y_1)^2 + (x_2 - y_2)^2 + \cdots + (x_n - y_n)^2]^{\frac{1}{2}}, \qquad (4.93)$$

the space is said to be a *Euclidean metric space of n-dimensions*; $(x_1, x_2, ..., x_n)$ are called the *rectangular cartesian coordinates* of P.

An orthogonal transformation of coordinates to a new system

$$(x'_1, x'_2, ..., x'_n)$$

with the same origin as the original system may be defined in much the same way as was done for a three-dimensional space in Chapter 1. The transformation matrix becomes

$$\begin{pmatrix} l_{11} & l_{12} & \cdots & l_{1n} \\ l_{21} & l_{22} & \cdots & l_{2n} \\ \vdots & & & \vdots \\ l_{n1} & l_{n2} & \cdots & l_{nn} \end{pmatrix} \qquad (4.94)$$

and is such that

$$l_{ki} l_{kj} = \delta_{ij} \qquad (4.95)$$

and

$$x'_i = l_{ij} x_j; \qquad (4.96)$$

observe that we are here using the summation convention and each repeated suffix ranges from 1 to n.

The definition of a vector given in § 2.2 and the definitions of addition, subtraction, etc., extend in a natural way. The n-dimensional analogue of (2.3) is

$$\mathbf{a} = (a_1, a_2, ..., a_n). \qquad (4.97)$$

The basic unit vectors of the n-dimensional system are

$$\begin{aligned} \mathbf{e}_1 &= (1, 0, 0, ..., 0) \\ \mathbf{e}_2 &= (0, 1, 0, ..., 0) \\ &\vdots \\ \mathbf{e}_n &= (0, 0, 0, ..., 1). \end{aligned} \qquad (4.98)$$

The scalar product of two vectors $\mathbf{a} = (a_1, a_2, ..., a_n)$ and $\mathbf{b} = (b_1, b_2, ..., b_n)$ is defined as

$$\mathbf{a} \cdot \mathbf{b} = a_1 b_1 + a_2 b_2 + \cdots + a_n b_n; \qquad (4.99)$$

this may be shown to be invariant under a translation or rotation of the axes, as in the three-dimensional case.

There is, however, one concept which does not extend to spaces of dimensionality greater than three, namely that of the vector product. In three-dimensions, there is just one unit vector normal to two given vectors (and associated with the given vectors by a right-hand screw rule), and it is essentially for this reason that it is possible to define a product of two vectors which is again a vector. When an attempt is made to generalise this to an n-dimensional space $(n > 3)$, it is found that there are $n-2$ unit vectors normal to two given vectors, and no analogue of the vector product exists.

Some of the definitions introduced in the present chapter extend quite readily to spaces of higher dimensions. For example, the gradient of a scalar field Ω, is defined as

$$\operatorname{grad}\Omega = \left(\frac{\partial\Omega}{\partial x_1}, \frac{\partial\Omega}{\partial x_2}, \ldots, \frac{\partial\Omega}{\partial x_n}\right). \tag{4.100}$$

Definition (4.13) of the directional derivative may also be extended in an obvious way, and it is easily shown that the relation (4.14) between the directional derivative and the gradient of a scalar field still holds. Further, the divergence of a vector field $\mathbf{F} = (F_1, F_2, \ldots, F_n)$ may be defined as

$$\operatorname{div}\mathbf{F} = \frac{\partial F_1}{\partial x_1} + \frac{\partial F_2}{\partial x_2} + \ldots + \frac{\partial F_n}{\partial x_n}. \tag{4.101}$$

However, the definition of curl $\mathbf{F}$ does *not* generalise, as may be expected in view of its close relation to the vector product.

Enough has perhaps now been said to indicate the lines upon which some parts of vector analysis may be generalised to spaces of n-dimensions. There are many other extensions less obvious than those mentioned here, but a full discussion lies well outside the scope of the present book.

EXERCISES

53. Define the Laplacian in an n-dimensional rectangular cartesian coordinate system, and verify that

$$\operatorname{div}(\operatorname{grad}\Omega) \equiv \nabla^2\Omega.$$

54. Prove the identity

$$\operatorname{div}(\phi\mathbf{F}) \equiv \phi\operatorname{div}\mathbf{F} + \mathbf{F}.\operatorname{grad}\phi$$

for an n-dimensional space.

CHAPTER 5

LINE, SURFACE, AND VOLUME INTEGRALS

In this chapter the concept of the Riemann integral of a function of a single real variable is extended to functions of three real variables.

5.1 Line integral of a scalar field

Let $\mathscr{C}$ be a piecewise smooth curve with intrinsic equation

$$\mathbf{r} = \mathbf{r}(s) = (x(s), y(s), z(s)), \quad 0 \leqslant s \leqslant l, \tag{5.1}$$

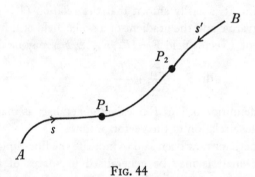

Fig. 44

and let $\Omega(x, y, z)$ be a scalar field defined at all points on $\mathscr{C}$. Then, on $\mathscr{C}$, Ω is a function of arc length s only. We define

$$I = \int_{s_1}^{s_2} \Omega(x(s), y(s), z(s)) \, ds, \tag{5.2}$$

where $0 \leqslant s_1 \leqslant s_2 \leqslant l$, to be *the line integral of Ω along the curve $\mathscr{C}$ from $s = s_1$ to $s = s_2$*, provided, of course, that the integral exists. An alternative notation is

$$I = \int_{P_1}^{P_2} \Omega(s) \, ds, \tag{5.3}$$

where P_1, P_2 are the points on $\mathscr{C}$ corresponding to s_1, s_2.

The line integral of a scalar field along a curve $\mathscr{C}$ is independent of the sense of description (or orientation) of $\mathscr{C}$.

Proof. Let A, B be the end points of $\mathscr{C}$, corresponding to $s=0$, $s=l$ respectively. Put

$$s' = l - s. \tag{5.4}$$

Then s' is arc length measured from B (Fig. 44). When s increases from 0 to l, the curve is described in the sense A to B; and when s' increases from 0 to l, the sense of description is from B to A.

Taking points P_1, P_2 on $\mathscr{C}$ corresponding to $s=s_1$, $s=s_2$,

$$\int_{P_1}^{P_2} \Omega\, ds = \int_{s_1}^{s_2} \Omega\, ds$$

$$= \int_{l-s_1}^{l-s_2} -\Omega\, ds', \quad \text{using (5.4)},$$

$$= \int_{l-s_2}^{l-s_1} \Omega\, ds'.$$

But $l-s_2$, $l-s_1$ are the values of s' at P_2, P_1, respectively. Hence

$$\int_{P_1}^{P_2} \Omega\, ds = \int_{P_2}^{P_1} \Omega\, ds',$$

showing that the line integral does not depend on the sense of description of $\mathscr{C}$.

EXAMPLE 1. The curve $\mathscr{C}$ is defined by the intrinsic equation

$$\mathbf{r} = (1, \sinh^{-1} s, (1+s^2)^{\frac{1}{2}}), \quad 0 \leqslant s \leqslant 1.$$

Evaluate the line integral of

$$\Omega = z^2 - x^2$$

along $\mathscr{C}$ from $s=0$ to $s=1$.

Solution. On $\mathscr{C}$, $\Omega=(1+s^2)-1=s^2$. Thus the required line integral is

$$\int_0^1 s^2\, ds = \tfrac{1}{3}.$$

Integral around a closed curve. Let $\mathscr{C}$ be a simple closed curve of total length l. The line integral

$$\int_0^l \Omega\, ds,$$

taken all the way round $\mathscr{C}$, is usually denoted by

$$\oint_{\mathscr{C}} \Omega\, ds. \tag{5.5}$$

We have already shown that the line integral of a scalar field along a curve is independent of the sense of description of the curve. *When the curve is closed, the line integral is also independent of the point from which arc length is measured.*

Proof. Let P, Q be different points on a closed curve $\mathscr{C}$. Let s denote arc length measured from P, and let s' denote arc length measured from Q (Fig. 45). Then at any point,

$$s = s' + \text{constant.} \tag{5.6}$$

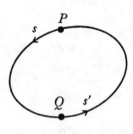

FIG. 45

Hence,

$$\oint_{\mathscr{C}} \Omega\, ds = \int_P^Q \Omega\, ds + \int_Q^P \Omega\, ds$$

$$= \int_P^Q \Omega\, ds' + \int_Q^P \Omega\, ds', \quad \text{using (5.6)}$$

$$= \oint_{\mathscr{C}} \Omega\, ds',$$

which proves the result stated.

EXAMPLE 2. Evaluate the line integral of

$$\Omega = (a^2 y^2/b^2 + b^2 x^2/a^2)^{\frac{1}{2}}$$

around the ellipse $\mathscr{C}$ with equation

$$x^2/a^2 + y^2/b^2 = 1, \quad z = 0.$$

Solution. The parametric equations of the ellipse can be taken as

$$x = a\cos\theta, \quad y = b\sin\theta, \quad z = 0, \quad 0 \leqslant \theta \leqslant 2\pi.$$

If s denotes arc length, we have

$$\frac{ds}{d\theta} = \left\{ \left(\frac{dx}{d\theta}\right)^2 + \left(\frac{dy}{d\theta}\right)^2 + \left(\frac{dz}{d\theta}\right)^2 \right\}^{\frac{1}{2}} = (a^2 \sin^2\theta + b^2 \cos^2\theta)^{\frac{1}{2}}.$$

Also, on $\mathscr{C}$,

$$\Omega = (a^2 \sin^2 \theta + b^2 \cos^2 \theta)^{\frac{1}{2}}.$$

Hence,

$$\oint_{\mathscr{C}} \Omega \, ds = \int_0^{2\pi} \Omega(\theta) \frac{ds}{d\theta} \, d\theta$$

$$= \int_0^{2\pi} (a^2 \sin^2 \theta + b^2 \cos^2 \theta) \, d\theta$$

$$= \pi(a^2 + b^2).$$

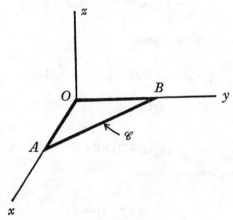

FIG. 46

EXAMPLE 3. Evaluate the line integral of

$$\Omega = x^2 + y^2$$

around the triangle $\mathscr{C}$ with vertices at the origin O and the points $A(1,0,0)$, $B(0,1,0)$ (Fig. 46).

Solution. The integral is evaluated in three parts.

(a) $$\int_O^A \Omega \, ds = \int_0^1 x^2 \, dx = \tfrac{1}{3}.$$

(b) $$\int_O^B \Omega \, ds = \int_0^1 y^2 \, dy = \tfrac{1}{3}.$$

(c) The equations of AB are

$$x + y = 1, \quad z = 0.$$

Hence, we can take

$$x = 1 - t, \quad y = t, \quad z = 0,$$

9

where t is a parameter which varies from 0 at A to 1 at B. If s denotes arc length of AB,

$$\frac{ds}{dt} = \left\{ \left(\frac{dx}{dt}\right)^2 + \left(\frac{dy}{dt}\right)^2 + \left(\frac{dz}{dt}\right)^2 \right\}^{\frac{1}{2}} = \sqrt{2}.$$

Also, on AB,

$$\Omega = (1-t)^2 + t^2.$$

Thus,

$$\int_A^B \Omega \, ds = \int_0^1 \Omega \frac{ds}{dt} dt$$

$$= \int_0^1 \{(1-t)^2 + t^2\}\sqrt{2} \, dt = \tfrac{2}{3}\sqrt{2}.$$

Combining (a), (b), and (c) now gives

$$\oint_{\mathscr{C}} \Omega \, ds = \int_O^A \Omega \, ds + \int_O^B \Omega \, ds + \int_A^B \Omega \, ds$$

$$= \tfrac{2}{3}(1+\sqrt{2}).$$

EXERCISES

1. Evaluate the line integral

$$\int_0^{2\pi a} (x+y+z) \, ds$$

along the curve with intrinsic equation

$$\mathbf{r} = (\tfrac{4}{5})^{\frac{1}{2}}(a \cos(s/2a), a \sin(s/2a), s).$$

2. A straight line is drawn parallel to the y-axis through the points $P(1,0,0)$, $Q(1,3,0)$. Show that

$$\int_P^Q (x^2+y^2+z^2) \, ds = 12.$$

[*Hint.* In this case $y=s$.]

3. Evaluate

$$\int_P^Q (x+y+z) \, ds$$

along the straight line joining the points $P(1,2,3)$ and $Q(4,5,6)$.

4. Evaluate

$$\oint_{\mathscr{C}} \Omega \, ds,$$

where $\Omega = y^2+z^2$ and $\mathscr{C}$ is the circle $x^2+y^2=a^2$, $z=0$.

5. A closed curve $\mathscr{C}$ is formed as follows: first, the origin O is joined to the point $A(1,0,0)$ by a straight line; then the point A is joined to the point $B(1,1,1)$ by part of the curve with parametric equation $\mathbf{r}=(1,t,t^2)$; finally, the point B is joined to the origin O by a straight line. Evaluate

$$\oint_{\mathscr{C}} xy\,ds.$$

5.2   Line integrals of a vector field

Scalar line integrals. Let the vector field $\mathbf{F}(x,y,z)$ be defined at all points on the piecewise smooth curve $\mathscr{C}$ given by equation (5.1). If $\hat{\mathbf{T}}$ denotes the unit tangent to $\mathscr{C}$, we define

$$I = \int_0^l \mathbf{F}.\hat{\mathbf{T}}\,ds \qquad (5.7)$$

to be the *scalar line integral* of $\mathbf{F}$ along $\mathscr{C}$ provided, of course, that the integral exists. Since $\hat{\mathbf{T}}=d\mathbf{r}/ds$, it is usual to put

$$\hat{\mathbf{T}}\,ds = d\mathbf{r}. \qquad (5.8)$$

Thus, (5.7) becomes

$$I = \int_0^l \mathbf{F}.d\mathbf{r}. \qquad (5.9)$$

If $\mathscr{C}$ is a simple closed curve, the line integral around $\mathscr{C}$ is denoted by

$$K = \oint_{\mathscr{C}} \mathbf{F}.d\mathbf{r}; \qquad (5.10)$$

K is called the *circulation* of $\mathbf{F}$ around $\mathscr{C}$.

When the sense of description of a curve is reversed, the direction of the unit tangent $\hat{\mathbf{T}}$ is also reversed. Thus, *the scalar line integral of a vector field along a curve changes sign when the sense of description of the curve is changed.*

EXAMPLE 4. Evaluate the scalar line integral of

$$\mathbf{F} = (z,x,y)$$

around the circle $x^2+y^2=0$, $z=0$, described in the clockwise sense relative to an observer looking along the positive z-axis.

Solution. The parametric equation of the circle $\mathscr{C}$ may be taken as

$$\mathbf{r} = (a\cos\theta, a\sin\theta, 0), \quad 0 \leqslant \theta \leqslant 2\pi,$$

the range of θ being chosen so that the sense of description is as required.
On $\mathscr{C}$,

$$\mathbf{F} = (0, a\cos\theta, a\sin\theta).$$

Also,

$$\frac{d\mathbf{r}}{d\theta} = (-a\sin\theta, a\cos\theta, 0).$$

Thus,

$$\oint_\mathscr{C} \mathbf{F}.d\mathbf{r} = \int_0^{2\pi} \mathbf{F}.\frac{d\mathbf{r}}{d\theta}\,d\theta$$

$$= \int_0^{2\pi} a^2\cos^2\theta\,d\theta$$

$$= \pi a^2.$$

EXAMPLE 5. Show that the circulation of any constant vector field **A** around any simple closed curve $\mathscr{C}$ is zero.

Solution. Choose the x-axis parallel to **A**. Then

$$\mathbf{A}.d\mathbf{r} = (A, 0, 0).(dx, dy, dz)$$

$$= A\,dx.$$

Thus

$$\oint_\mathscr{C} \mathbf{A}.d\mathbf{r} = A\oint_\mathscr{C} dx.$$

If P_1, P_2 are points on $\mathscr{C}$ corresponding to $x=x_1$, $x=x_2$ respectively, then integrating along $\mathscr{C}$,

$$\int_{P_1}^{P_2} dx = x_2-x_1.$$

Allowing the point P_2 to move around $\mathscr{C}$ until it coincides with P_1, we have $x_2=x_1$, and hence

$$\oint_\mathscr{C} dx = \int_{P_1}^{P_2} dx = 0.$$

The result required follows at once.

Vector line integrals. If the components F_1, F_2, F_3 of the vector field

$$\mathbf{F} = F_1\mathbf{i}+F_2\mathbf{j}+F_3\mathbf{k}$$

are integrable along a curve $\mathscr{C}$, the *vector line integral of* **F** *along* $\mathscr{C}$, from $s=s_1$ to $s=s_2$, is defined as

$$\int_{s_1}^{s_2} \mathbf{F}\,ds = \mathbf{i}\int_{s_1}^{s_2} F_1\,ds+\mathbf{j}\int_{s_1}^{s_2} F_2\,ds+\mathbf{k}\int_{s_1}^{s_2} F_3\,ds. \tag{5.11}$$

The unit tangent $\hat{\mathbf{T}}$ to a curve $\mathscr{C}$ is itself a vector field defined on $\mathscr{C}$. Thus, we may use (5.11) to evaluate a *vector line integral of a scalar field* Ω, which is defined as

$$\int \Omega\,d\mathbf{r} = \int \Omega\hat{\mathbf{T}}\,ds. \tag{5.12}$$

It is clear that $\hat{\mathbf{T}}$ may also appear as part of a vector product, leading to line integrals of the form

$$\int \mathbf{F} \times d\mathbf{r} = -\int d\mathbf{r} \times \mathbf{F} = \int \mathbf{F} \times \hat{\mathbf{T}} \, ds. \tag{5.13}$$

The reader should experience little difficulty in working with these new forms, as they appear in the following exercises.

EXERCISES

6. If $\mathbf{F} = (x, 2y, 3z)$, evaluate

$$\text{(i)} \int_O^A \mathbf{F} . d\mathbf{r}, \quad \text{(ii)} \int_O^A \mathbf{F} \times d\mathbf{r}, \quad \text{(iii)} \int_O^A \mathbf{F} \, ds,$$

along the curve $\mathbf{r} = (t, \sqrt{\tfrac{1}{2}}t^2, \tfrac{1}{3}t^3)$ from the origin O to the point $A(1, \tfrac{1}{2}\sqrt{2}, \tfrac{1}{3})$.

7. Evaluate

$$\int z \, d\mathbf{r}$$

along the curve $\mathbf{r} = (a\cos t, b\sin t, ct)$ from the point $t = 0$ to the point $t = 2\pi$.

8. If $\mathbf{F} = (y\mathbf{i} - x\mathbf{j})/(x^2 + y^2)$, and $\mathscr{C}$ is the circle $x^2 + y^2 = a^2$ in the xy-plane described in the anticlockwise sense, evaluate

$$\oint_{\mathscr{C}} \mathbf{F} . d\mathbf{r}.$$

9. Evaluate

$$\oint_{\mathscr{C}} \mathbf{F} . d\mathbf{r},$$

where $\mathbf{F} = (x - 3y, y - 2x, 0)$ and $\mathscr{C}$ is the perimeter of the ellipse $x^2/9 + y^2/4 = 1$ in the xy-plane, described in the anticlockwise sense.

10. Evaluate

$$\int_{\mathscr{C}} \mathbf{r} \times d\mathbf{r} \quad \text{and} \quad \int_{\mathscr{C}} \mathbf{r} \, ds$$

from the point $(a, 0, 0)$ to the point $(a, 0, 2\pi b)$ on the circular helix

$$\mathbf{r} = (a\cos t, a\sin t, bt).$$

5.3 Repeated integrals

Before discussing surface and volume integrals, it is necessary to introduce the idea of repeated integration and (in the next section) double and triple integrals.

Let $p(x)$, $q(x)$ be functions of x, and consider the following integral of a function $f(x,y)$:

$$I(x) = \int_{p(x)}^{q(x)} f(x,y)\,dy.$$

Now integrate $I(x)$ between the limits $x=a$, $x=b$. We write

$$\int_a^b \left\{ \int_{p(x)}^{q(x)} f(x,y)\,dy \right\} dx = \int_a^b \int_{p(x)}^{q(x)} f(x,y)\,dy\,dx. \tag{5.14}$$

Such integrals are called *repeated integrals*.

Two important properties are:

(i) If $p=c$, $q=d$, where c, d are constants, then

$$\int_a^b \int_c^d f(x,y)\,dy\,dx = \int_c^d \int_a^b f(x,y)\,dx\,dy; \tag{5.15}$$

i.e. the order of integration is immaterial when the limits of integration are constants.

(ii) If $f(x,y) = \phi(x)\psi(y)$, then

$$\int_a^b \int_c^d \phi(x)\,\psi(y)\,dy\,dx = \left(\int_a^b \phi(x)\,dx \right)\left(\int_c^d \psi(y)\,dy \right); \tag{5.16}$$

i.e. when the integrand is a *separable* function of x and y, and the limits of integration are constants, a repeated integral separates into the product of two single integrals. The first property is proved in the next section. The second property may be proved by observing that the first integration (with respect to y) does not affect $\phi(x)$.

EXAMPLE 6. Evaluate

$$I = \int_0^1 \int_{\frac{1}{4}x}^{\frac{1}{2}} xy\,dy\,dx.$$

Solution. Performing the integration with respect to y, we have

$$I = \int_0^1 [\tfrac{1}{2}xy^2]_{y=\frac{1}{4}x}^{y=\frac{1}{2}}\,dx$$

$$= \int_0^1 (\tfrac{1}{8}x - \tfrac{1}{8}x^3)\,dx$$

$$= \tfrac{1}{32}.$$

EXAMPLE 7. Evaluate

$$I = \int_0^1 \int_0^2 (x^2+y)\,dy\,dx,$$

and verify that the same result is obtained when the order of integration is inverted.

Solution. We have

$$I = \int_0^1 [x^2y+\tfrac{1}{2}y^2]_{y=0}^{y=2}\,dx$$

$$= \int_0^1 (2x^2+2)\,dx = \tfrac{8}{3}.$$

Inverting the order of integration,

$$I = \int_0^2 \int_0^1 (x^2+y)\,dx\,dy$$

$$= \int_0^2 [\tfrac{1}{3}x^3+xy]_{x=0}^{x=1}\,dy$$

$$= \int_0^2 (\tfrac{1}{3}+y)\,dy = \tfrac{8}{3}, \quad \text{as before.}$$

EXERCISES

11. Show that

$$\int_0^{\frac{1}{2}\pi} \int_0^{\frac{1}{3}\pi} \sin x \cos y \,dx\,dy = 1-\tfrac{1}{2}\sqrt{3}.$$

12. Show that

$$\int_0^1 \int_0^{1-x} (x+y)^2 \,dy\,dx = \tfrac{1}{4}.$$

13. Show that

$$\int_0^\pi \int_0^{\sin y} dx\,dy = 2.$$

5.4 Double and triple integrals

The Riemann integral. We first recall the following definition of the Riemann integral of a function $f(x)$.

Let $f(x)$ be defined in the interval $a \leqslant x \leqslant b$. Subdivide the interval into m sub-intervals of lengths $\delta x_1, \delta x_2, \ldots, \delta x_m$, and let $x_1, x_2, \ldots, x_m$ be any points within the respective sub-intervals. Let the method of subdivision be such that $\delta x_1, \delta x_2, \ldots, \delta x_m$ are all less than some quantity ϵ_m such that

$$\epsilon_m \to 0 \quad \text{as} \quad m \to \infty. \tag{5.17}$$

Then the Riemann integral of $f(x)$ over the interval $a \leqslant x \leqslant b$ is defined as

$$\lim_{m \to \infty} \sum_{r=1}^{m} f(x_r)\,\delta x_r = \int_{a}^{b} f(x)\,dx, \tag{5.18}$$

provided that the limit exists and is the same for all methods of subdivision satisfying (5.17).

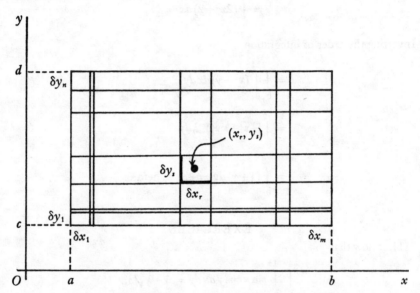

FIG. 47. Method of subdivision for double integrals

Double integrals. Let $f(x,y)$ be defined in the rectangle $a \leqslant x \leqslant b$, $c \leqslant y \leqslant d$. Subdivide the interval $a \leqslant x \leqslant b$ into m sub-intervals of lengths $\delta x_1, \delta x_2, \ldots,$ δx_m; and subdivide the interval $c \leqslant y \leqslant d$ into n sub-intervals of lengths $\delta y_1, \delta y_2, \ldots, \delta y_n$. Consider the sum

$$S_{mn} = \sum_{r=1}^{m} \sum_{s=1}^{n} f(x_r, y_s)\,\delta x_r\,\delta y_s, \tag{5.19}$$

where (x_r, y_s) is any point inside the sub-rectangle of area $\delta x_r\,\delta y_s$ (Fig. 47). Choose any method of subdivision such that $\delta x_1, \ldots, \delta x_m$ are all less than ϵ_m and $\delta y_1, \ldots, \delta y_n$ are all less than η_n, where

$$\left.\begin{array}{l} \epsilon_m \to 0 \quad \text{as} \quad m \to \infty \\ \eta_n \to 0 \quad \text{as} \quad n \to \infty. \end{array}\right\} \tag{5.20}$$

We then define

$$\lim_{\substack{m \to \infty \\ n \to \infty}} S_{mn} = \iint_{\mathscr{R}} f(x,y)\, dx\, dy \qquad (5.21)$$

as the double integral of $f(x,y)$, where $\mathscr{R}$ denotes the rectangle $a \leqslant x \leqslant b$, $c \leqslant y \leqslant d$, provided the limit exists and is the same for all methods of subdivision satisfying (5.20).

The double integral defined in this way exists in nearly all cases of practical interest.

The double integral of $f(x,y)$ over a closed region $\mathscr{R}$ which is not rect-

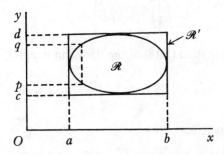

FIG. 48. Double integration over a non-rectangular region

angular is defined as follows. Let $\mathscr{R}'$ be a rectangle, with sides parallel to the x-, y-axes, enclosing $\mathscr{R}$ (Fig. 48). Define

$$g(x,y) = \begin{cases} f(x,y) & \text{if } (x,y) \text{ is in } \mathscr{R} \\ 0 & \text{if } (x,y) \text{ is not in } \mathscr{R}. \end{cases} \qquad (5.22)$$

Then we define

$$\iint_{\mathscr{R}} f(x,y)\, dx\, dy = \iint_{\mathscr{R}'} g(x,y)\, dx\, dy; \qquad (5.23)$$

where the latter integral, being over a rectangular region, is as defined by (5.21).

Evaluation of double integrals. A double integral is usually evaluated by expressing it as a repeated integral. A detailed proof that a double integral can be expressed as a repeated integral lies outside the scope of this book.[1] However, the following argument (which is by no means rigorous) shows that the result is plausible.

[1] See, for example, R. Courant: *Differential and Integral Calculus* (Blackie), Vol. II, pp. 236–239.

Consider first the case when $f(x,y)$ is defined over the rectangle $a \leqslant x \leqslant b$, $c \leqslant y \leqslant d$. Then

$$\iint\limits_{\mathscr{R}} f(x, y)\, dx\, dy = \lim_{m \to \infty} \lim_{n \to \infty} \sum_{r=1}^{m} \sum_{s=1}^{n} f(x_r, y_s)\, \delta x_r \delta y_s$$

$$= \lim_{n \to \infty} \sum_{s=1}^{n} \left\{ \lim_{m \to \infty} \sum_{r=1}^{m} f(x_r, y_s)\, \delta x_r \right\} \delta y_s$$

$$= \lim_{n \to \infty} \sum_{s=1}^{n} \left\{ \int_a^b f(x, y_s)\, dx \right\} \delta y_s, \quad \text{using (5.18)},$$

$$= \int_c^d \left\{ \int_a^b f(x, y)\, dx \right\} dy,$$

again using the definition (5.18) of the Riemann integral. Thus

$$\iint\limits_{\mathscr{R}} f(x, y)\, dx\, dy = \int_c^d \int_a^b f(x, y)\, dx\, dy. \tag{5.24}$$

A similar argument gives

$$\iint\limits_{\mathscr{R}} f(x, y)\, dx\, dy = \int_a^b \int_c^d f(x, y)\, dy\, dx. \tag{5.25}$$

We note that, combining (5.24) and (5.25), (5.15) follows. Thus the order of integration in the repeated integral is immaterial provided the corresponding double integral exists.

When $\mathscr{R}$ is not a rectangular region, enclose it by the rectangle $a \leqslant x \leqslant b$, $c \leqslant y \leqslant d$, as shown in Fig. 48. Assume that any line parallel to Oy cuts $\mathscr{R}$ in at most two points (the extension to more general cases is easy and is left to the reader). Then, by the definition (5.23) and the result (5.25),

$$I = \iint\limits_{\mathscr{R}} f(x, y)\, dy\, dx = \int_a^b \int_c^d g(x, y)\, dy\, dx.$$

At a fixed value of x, let $p(x)$, $q(x)$ be the extreme values of y in $\mathscr{R}$ (Fig. 48). Then, since $g(x, y) = f(x, y)$ in the range $p(x) \leqslant y \leqslant q(x)$ and $g(x, y) = 0$ in the ranges $c \leqslant y \leqslant p(x)$ and $q(x) \leqslant y \leqslant d$,

$$I = \int_a^b \int_{p(x)}^{q(x)} f(x, y)\, dy\, dx. \tag{5.26}$$

If the order of integration is inverted, the limits of the inner integral will be functions of y, and the limits of the outer integral will be c, d.

EXAMPLE 8. Express

$$I = \int\limits_0^1 \int\limits_{\frac{1}{2}x}^{\frac{1}{2}} xy \, dy \, dx$$

as a double integral. Evaluate I by inverting the order of integration.

Solution. We observe that x ranges from 0 to 1, and that, for any x, the extreme values of y are $\frac{1}{2}x$ and $\frac{1}{2}$. Thus the triangular region $\mathscr{R}$, shown in Fig. 49, is covered. It follows that, expressed as a double integral,

$$I = \iint\limits_{\mathscr{R}} xy \, dx \, dy.$$

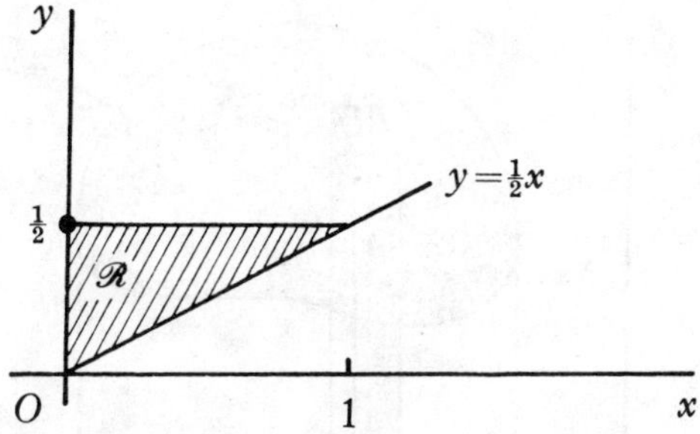

FIG. 49

Now $\mathscr{R}$ is also covered if y ranges from 0 to $\frac{1}{2}$ and, for every value of y, x ranges from 0 to $2y$. Thus

$$I = \int\limits_0^{\frac{1}{2}} \int\limits_0^{2y} xy \, dx \, dy$$

$$= \int\limits_0^{\frac{1}{2}} [\tfrac{1}{2}x^2 y]_{x=0}^{x=2y} \, dy$$

$$= \int\limits_0^{\frac{1}{2}} 2y^3 \, dy$$

$$= \tfrac{1}{32}.$$

We note that this integral was evaluated in Example 6, p. 124, by integrating first with respect to y and then with respect to x; the two results obtained agree.

Geometrical interpretation of double integrals. In rectangular cartesian coordinates x, y, z, the equation

$$z = f(x, y) \tag{5.27}$$

represents a surface (we deal with surfaces in more detail in § 5.5).

Let $f(x,y) > 0$ for all points (x,y) in a region $\mathscr{R}$ of the xy-plane. From a rectangular element of $\mathscr{R}$, of area $\delta x_r \delta y_s$, erect perpendiculars from the xy-plane to meet the surface, as shown in Fig. 50. Then, if (x_r, y_s) is any point in the rectangle, the volume of the columnar element so formed is approximately $f(x_r, y_s)\delta x_r \delta y_s$. As the double integral

$$\iint\limits_{\mathscr{R}} f(x, y)\, dx\, dy$$

is the limit of a sum of terms of this form, it is intuitively evident that the double integral represents the volume 'underneath the surface'; that is, the

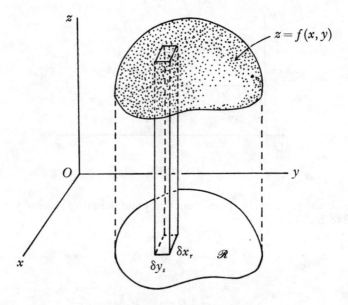

FIG. 50. Volume under a surface

volume bounded by the surface, its projection $\mathscr{R}$ on the xy-plane and the cylinder with generators through the boundary of $\mathscr{R}$ and parallel to Oz.

Similar reasoning also leads to the conclusion that the double integral

$$\iint\limits_{\mathscr{R}} dx\, dy$$

represents the area of the region $\mathscr{R}$.

Change of variables. Evaluation of a double integral is often made easier by a suitable change of variables.

If variables u, v are such that

$$x = x(u, v), \quad y = y(u, v), \tag{5.28}$$

it can be shown[1] that

$$\iint_{\mathscr{R}} f(x, y) \, dx \, dy = \iint_{\mathscr{R}'} f\{x(u, v), y(u, v)\} \, |J| \, du \, dv, \qquad (5.29)$$

where $\mathscr{R}'$ is the region in the uv-plane consisting of points corresponding to the points of $\mathscr{R}$ in the xy-plane, and

$$J = \frac{\partial(x, y)}{\partial(u, v)} = \begin{vmatrix} \dfrac{\partial x}{\partial u} & \dfrac{\partial x}{\partial v} \\[2mm] \dfrac{\partial y}{\partial u} & \dfrac{\partial y}{\partial v} \end{vmatrix}. \qquad (5.30)$$

The determinant J is called the *Jacobian of the transformation* (5.28). Various restrictions must be placed on the functions $x(u,v)$, $y(u,v)$, but we shall omit the details.

A very common change of variables is from rectangular cartesian coordinates x, y to plane polar coordinates r, θ, defined by

$$x = r \cos \theta, \quad y = r \sin \theta. \qquad (5.31)$$

The Jacobian of this transformation is

$$\frac{\partial(x, y)}{\partial(r, \theta)} = \begin{vmatrix} \cos \theta & -r \sin \theta \\ \sin \theta & r \cos \theta \end{vmatrix} = r.$$

Thus,

$$\iint_{\mathscr{R}} f(x, y) \, dx \, dy = \iint_{\mathscr{R}'} f(r \cos \theta, r \sin \theta) \, r \, dr \, d\theta, \qquad (5.32)$$

where $\mathscr{R}'$ is the region in the $r\theta$-plane corresponding to the region $\mathscr{R}$ in the xy-plane. The plausibility of this result can be seen as follows.

When a double integral is defined using rectangular cartesian coordinates, the region of integration is subdivided into elementary regions by lines $x = \text{constant}$, $y = \text{constant}$; the area of a typical element may be denoted by $\delta x \delta y$. In the case of polar coordinates, the corresponding subdivision is into elements bounded by circles $r = \text{constant}$ and lines $\theta = \text{constant}$ (Fig. 51(a)). The area of a typical element is approximately $r \delta r \delta \theta$. Recalling the definition (5.21), we should thus expect the double integral of $f(x,y)$ to transform to the limit of a sum of terms of the form $f(r \cos \theta, r \sin \theta) r \delta r \delta \theta$, and this leads to the result (5.32).

EXAMPLE 9. Evaluate

$$I = \iint_{\mathscr{R}} e^{-(x^2 + y^2)} \, dx \, dy,$$

[1] See, for example, R. Courant: *Differential and Integral Calculus* (Blackie), Vol. II, pp. 247–253.

where $\mathscr{R}$ denotes the quadrant $x \geqslant 0$, $y \geqslant 0$ of the xy-plane. Deduce that

$$\int_0^\infty e^{-x^2}\,dx = \tfrac{1}{2}\sqrt{\pi}.$$

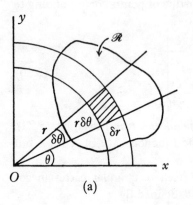

(a)

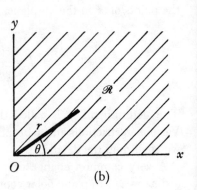

(b)

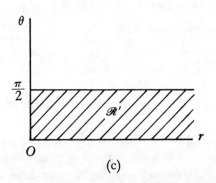

(c)

FIG. 51. Use of plane polar coordinates to evaluate double integrals: (a) the element $r\delta r\delta\theta$; (b) region of integration for Example 9, in xy-plane; (c) region of integration for Example 9, in $r\theta$-plane

Solution. Substituting $x = r\cos\theta$, $y = r\sin\theta$, the integral becomes

$$I = \iint_{\mathscr{R}'} e^{-r^2} r\,dr\,d\theta.$$

All points of the region $\mathscr{R}$ are covered if θ ranges from 0 to $\tfrac{1}{2}\pi$ and if, for every value of θ in this range, r ranges from 0 to ∞ (Fig. 51(b)). Thus the region $\mathscr{R}'$ in the $r\theta$-plane is the semi-infinite strip $0 \leqslant r < \infty$, $0 \leqslant \theta \leqslant \tfrac{1}{2}\pi$ (Fig. 51(c)). It follows that

$$I = \int_0^{\frac{1}{2}\pi} \int_0^\infty r\,e^{-r^2}\,dr\,d\theta.$$

Observing that

$$\int_0^\infty r \exp(-r^2)\, dr = -[\tfrac{1}{2}\exp(-r^2)]_0^\infty = \tfrac{1}{2}$$

gives

$$I = \tfrac{1}{4}\pi.$$

In terms of rectangular coordinates,

$$I = \int_0^\infty \int_0^\infty e^{-(x^2+y^2)}\, dx\, dy.$$

Using (5.16) gives

$$I = \left(\int_0^\infty e^{-x^2}\, dx\right)\left(\int_0^\infty e^{-y^2}\, dy\right)$$

$$= \left(\int_0^\infty e^{-x^2}\, dx\right)^2.$$

Hence

$$\int_0^\infty e^{-x^2}\, dx = \tfrac{1}{2}\sqrt{\pi},$$

the positive root being chosen since the result is clearly positive.

EXAMPLE 10. Evaluate

$$I = \iint_{\mathscr{R}} (1 - x^2/a^2 - y^2/b^2)^{\frac{1}{2}}\, dx\, dy,$$

where $\mathscr{R}$ is the region bounded by the ellipse $(x^2/a^2)+(y^2/b^2)=1$. Hence show that the volume of the ellipsoid

$$x^2/a^2 + y^2/b^2 + z^2/c^2 = 1$$

is $\tfrac{4}{3}\pi abc$.

Solution. Consider the transformation to variables r, θ (which are *not* in this case polar coordinates) defined by

$$x = ar\cos\theta, \quad y = br\sin\theta.$$

It is easily verified that the Jacobian of this transformation is abr, and also that the region bounded by the ellipse is covered if r ranges from 0 to 1 and θ ranges from 0 to 2π. Hence we find

$$I = \int_0^1 \int_0^{2\pi} abr(1-r^2)^{\frac{1}{2}}\, dr\, d\theta$$

which, after a short calculation, becomes

$$I = \tfrac{2}{3}\pi ab.$$

The volume of the given ellipsoid (the upper portion of which is shown in Fig. 52) is

$$V = 2 \iint_{\mathscr{R}} z \, dx \, dy,$$

where $z = c\{1 - (x^2/a^2) - (y^2/b^2)\}^{\frac{1}{2}}$ and $\mathscr{R}$ is the region in the xy-plane bounded by the ellipse $(x^2/a^2) + (y^2/b^2) = 1$. Hence

$$V = 2cI = \tfrac{4}{3}\pi abc,$$

as required.

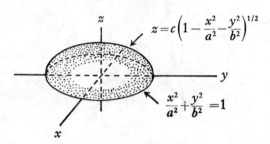

FIG. 52

Triple integrals. The definition of a double integral can be extended quite easily to define the triple integral of a function $f(x, y, z)$ over a closed region τ. The integral is denoted thus:

$$\iiint_{\tau} f(x, y, z) \, dx \, dy \, dz.$$

Like double integrals, triple integrals are usually evaluated by repeated integration. The method is similar to that used to evaluate double integrals, and it will be sufficient to give examples.

EXAMPLE 11. If τ denotes the region

$$|x| \leqslant a, \quad |y| \leqslant b, \quad |z| \leqslant c,$$

evaluate

$$I = \iiint_{\tau} (y^2 + z^2) \, dx \, dy \, dz.$$

Solution. The region τ is a rectangular parallelepiped with faces in the planes $x = \pm a, y = \pm b, z = \pm c$. Thus

$$I = \int_{-c}^{c} \int_{-b}^{b} \int_{-a}^{a} (y^2 + z^2) \, dx \, dy \, dz$$

$$= \int_{-c}^{c} \int_{-b}^{b} 2a(y^2 + z^2) \, dy \, dz \qquad \text{(integrating with respect to } x\text{)}.$$

Hence, $I = \int\limits_{-c}^{c} 2a\left(\dfrac{2b^3}{3}+2bz^2\right)dz$ (integrating with respect to y)

$= \tfrac{8}{3}abc(b^2+c^2)$.

EXAMPLE 12. A tetrahedron has vertices at the points $O(0,0,0)$, $A(1,0,0)$, $B(0,1,0)$, $C(0,0,1)$. If τ denotes the region bounded by the tetrahedron, evaluate

$$I = \iiint\limits_{\tau} z\,dx\,dy\,dz.$$

Solution. Let P be any point $(x,y,0)$ inside the triangle OAB, and let the line through P parallel to Oz meet the plane ABC at Q (Fig. 53). The equation of the plane ABC is

$$x+y+z = 1.$$

Thus at Q, $z=1-x-y$.

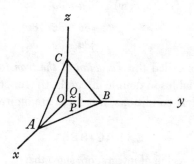

FIG. 53

To cover the volume of integration, we let x, y range over all points P in the triangle OAB, and let z range from zero (its value at P) to $1-x-y$ (its value at Q). Thus

$$I = \iint\limits_{\triangle OAB}\left(\int\limits_{0}^{1-x-y} z\,dz\right)dx\,dy$$

$$= \iint\limits_{\triangle OAB} \tfrac{1}{2}(1-x-y)^2\,dx\,dy$$

$$= \int\limits_{0}^{1}\int\limits_{0}^{1-y} \tfrac{1}{2}(1-x-y)^2\,dx\,dy$$

$$= \int\limits_{0}^{1} [-\tfrac{1}{6}(1-x-y)^3]_{x=0}^{x=1-y}\,dy$$

$$= \int\limits_{0}^{1} \tfrac{1}{6}(1-y)^3\,dy$$

$$= \tfrac{1}{24}.$$

Change of variables in triple integrals. If variables u, v, w are defined by the relations

$$x = x(u, v, w), \quad y = y(u, v, w), \quad z = z(u, v, w), \qquad (5.33)$$

then it can be shown that

$$\iiint_\tau f(x, y, z) \, dx \, dy \, dz = \iiint_{\tau'} f\{x(u, v, w), y(u, v, w), z(u, v, w)\}$$

$$\times |J| \, du \, dv \, dw, \qquad (5.34)$$

where τ' is the region in the uvw-space corresponding to the region τ in the xyz-space, and

$$J = \frac{\partial(x, y, z)}{\partial(u, v, w)} = \begin{vmatrix} \dfrac{\partial x}{\partial u} & \dfrac{\partial x}{\partial v} & \dfrac{\partial x}{\partial w} \\[2mm] \dfrac{\partial y}{\partial u} & \dfrac{\partial y}{\partial v} & \dfrac{\partial y}{\partial w} \\[2mm] \dfrac{\partial z}{\partial u} & \dfrac{\partial z}{\partial v} & \dfrac{\partial z}{\partial w} \end{vmatrix}. \qquad (5.35)$$

The determinant J is called the *Jacobian of the transformation*. As in the case of changes of variables in double integrals, the functions involved must satisfy certain analytical conditions, but these are omitted here.

EXERCISES

14. By sketching the region of integration, show that

$$\int_0^1 \int_{\sqrt{y}}^1 dx \, dy = \int_0^1 \int_0^{x^2} dy \, dx.$$

Evaluate the integrals.

15. By inverting the order of integration, evaluate

$$\int_0^1 \int_{y^2}^{y^{1/2}} (y/x) \, e^x \, dx \, dy.$$

16. A region $\mathscr{R}$ is bounded by a triangle with vertices at the origin and the points $(1, 1)$, $(-1, 1)$. Show that

$$\iint_{\mathscr{R}} e^{y^2} \, dx \, dy = (e-1).$$

17. Evaluate

$$\int_0^1 \int_{\frac{1}{2}y}^y \frac{xy^2}{\sqrt{(x^3+y^3)}} \, dx \, dy + \int_1^2 \int_{\frac{1}{2}y}^1 \frac{xy^2}{\sqrt{(x^3+y^3)}} \, dx \, dy.$$

[*Hint.* Sketch the regions of integration.]

18. Evaluate

$$\iint\limits_{\mathscr{R}} (x^2+y^2)\, dx\, dy,$$

where $\mathscr{R}$ is the region bounded by the circles $x^2+y^2=a^2$, $x^2+y^2=b^2$, $(a<b)$. [*Hint.* Use plane polar coordinates r, θ.]

19. Find the volume bounded by the paraboloid

$$z = 4-x^2-y^2$$

and the xy-plane.

20. By transforming to plane polar coordinates r, θ, find

$$\int\limits_{0}^{\frac{1}{2}\sqrt{2}} \int\limits_{x}^{\sqrt{(1-x^2)}} \frac{\log(x^2+y^2)}{\sqrt{(x^2+y^2)}}\, dy\, dx.$$

21. By means of the transformation $x=ar\cos\theta$, $y=br\sin\theta$, evaluate

$$\iint\limits_{\mathscr{R}} x^2\, dx\, dy,$$

where $\mathscr{R}$ denotes the region bounded by the ellipse $x^2/a^2+y^2/b^2=1$.

22. By means of the substitution $x=(r\cos\theta)^{\frac{1}{2}}$, $y=(r\sin\theta)^{\frac{1}{2}}$, prove that

$$\iint\limits_{\mathscr{R}} x^3(1-x^4-y^4)\, dx\, dy = 4/45,$$

where $\mathscr{R}$ is the region defined by $x\geqslant 0$, $y\geqslant 0$, $x^4+y^4\leqslant 1$.

23. Find

$$\int\limits_{0}^{1} \int\limits_{0}^{1} \int\limits_{0}^{y} xyz\, dx\, dy\, dz.$$

24. Evaluate

$$\iiint\limits_{\tau} \exp(-x^2-y^2-z^2)\, dx\, dy\, dz,$$

where τ denotes the whole of space.

25. The vertices of a triangular prism are the points

$$(0,0,0), (1,0,0), (0,1,0), (0,0,2), (0,1,2), (1,0,2).$$

Evaluate

$$\iiint\limits_{\tau} x^2\, dx\, dy\, dz,$$

where τ is the region bounded by the prism.

26. If the variables x, y, z are related to the variables r, θ, ϕ by the equations

$$x = ar\sin\theta\cos\phi, \quad y = br\sin\theta\sin\phi, \quad z = cr\cos\theta,$$

show that

$$\frac{\partial(x, y, z)}{\partial(r, \theta, \phi)} = abcr^2 \sin \theta.$$

By means of the above transformation of variables, evaluate

$$\iiint_\tau (x^2 + y^2 + z^2)\, dx\, dy\, dz,$$

where τ denotes the region bounded by the ellipsoid

$$x^2/a^2 + y^2/b^2 + z^2/c^2 = 1.$$

5.5   Surfaces

Let a variable point P have position vector

$$\overrightarrow{OP} = \mathbf{r}(u, v) = \{x(u, v), y(u, v), z(u, v)\}, \qquad (5.36)$$

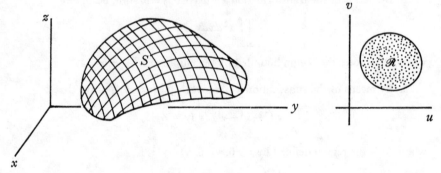

Fig. 54

where: (i) u, v are continuous parameters taking all values in a region $\mathcal{R}$ of the uv-plane; (ii) x, y, and z are continuous, single-valued functions of u, v in $\mathcal{R}$.

If v is given fixed values, in turn, and in each case u is allowed to vary over its range of values, P will trace out a family of curves called *u-coordinate curves*. Similarly, when fixed values are assigned to u and v is varied, the loci of P are the family of *v-coordinate curves* (Fig. 54). The network of all these curves forms a surface S, say, which is thus the locus of P.

Equation (5.36) defines a *mapping* of the region $\mathcal{R}$ in the uv-plane onto the surface S in the xyz-space (Fig. 54). Since x, y and z are single-valued functions of u and v, each point in $\mathcal{R}$ will correspond to just one point on S. However, it would be too restrictive for our purposes to consider only mappings which are such that each point on S also corresponds to just one point of $\mathcal{R}$ (that is, one–one mappings). The class of mappings considered will be widened to include those which are such that most, but not necessarily all, points on S correspond to one point of $\mathscr{R}$: more precisely, we include mappings with *exceptional points* on S, each corresponding to more than one point of $\mathscr{R}$, provided that these exceptional points are either isolated from other exceptional points or (at most) constitute a finite number of curves on S. It will be found that such exceptional points cause no special difficulty.

EXAMPLE 13. Consider

$$\overrightarrow{OP} = \mathbf{r}(\phi, z) = (a\cos\phi, a\sin\phi, z), \tag{5.37}$$

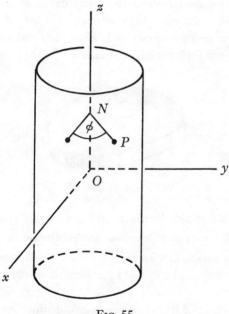

FIG. 55

where $0 \leqslant \phi \leqslant 2\pi$, $-\infty < z < \infty$, and a is a constant. If z takes a fixed value z_0, then since $x = a\cos\phi$, $y = a\sin\phi$, P describes a circle in the plane $z = z_0$ as ϕ ranges from 0 to 2π. If ϕ is fixed and z varies, P describes a line parallel to Oz and at a distance a from it. The complete locus of P is thus an infinite circular cylinder of radius a with axis Oz (Fig. 55). We note that each of the points $(a, 0, z)$ on the generator which cuts the positive x-axis corresponds to the two points $(0, z)$ and $(2\pi, z)$ in the ϕz-plane. Thus this generator consists of exceptional points of the mapping (5.37).

EXAMPLE 14. If

$$\overrightarrow{OP} = \mathbf{r}(\theta, \phi) = (a\sin\theta\cos\phi, a\sin\theta\sin\phi, a\cos\theta), \tag{5.38}$$

where $0 \leqslant \theta \leqslant \pi$, $0 \leqslant \phi \leqslant 2\pi$, then

$$x^2 + y^2 + z^2 = a^2 \sin^2 \theta (\cos^2 \phi + \sin^2 \phi) + a^2 \cos^2 \theta = a^2.$$

In this case P lies on a sphere of radius a, centred at the origin, for all values of θ and ϕ, and it may be verified that with the ranges of θ and ϕ given, the whole sphere is covered. It is left as an exercise for the reader to verify that the semicircle $0 \leqslant \theta \leqslant \pi$, $\phi = 0$ consists of exceptional points of the mapping (5.38).

Two-sided surfaces. The reader may be familiar with the fact that a surface may have only one side. For example, if a strip of paper is twisted once and the ends glued together, a one-sided surface, called a Möbius strip (Fig. 56) is obtained. It can be verified that this surface has only one side by tracing a continuous line along the strip and observing that all points can be reached without crossing over the edge. Such surfaces will henceforth be excluded from consideration, and we discuss only two-sided surfaces.

It will be sufficient for our purposes to rely upon the intuitive notion of

FIG. 56. A Möbius strip

what is meant by a two-sided surface, and accordingly an analytical definition is omitted.

Open and closed surfaces. A surface S is said to be *open* if every pair of points not lying on S can be joined by a continuous curve which does not cross S.

A surface S is *closed* if it divides space into distinct regions, $\mathscr{R}_1$ and $\mathscr{R}_2$ say, such that every continuous curve joining a point in $\mathscr{R}_1$ to a point in $\mathscr{R}_2$ crosses S at least once.

For example, a cap of a sphere is an open surface, but the complete surface of a sphere is closed.

Unit normal vector. Consider the surface S defined by (5.36). As we have seen, the locus of P when u varies and v is fixed is a u-coordinate curve on S, and when v varies and u is fixed the locus is a v-coordinate curve. At any point P on S, the vectors

$$\frac{\partial \mathbf{r}}{\partial u} = \mathbf{r}_u, \quad \frac{\partial \mathbf{r}}{\partial v} = \mathbf{r}_v \tag{5.39}$$

are respectively tangential to the u-, v-coordinate curves through P (Fig. 57).

The vector

$$\mathbf{n} = \frac{\mathbf{r}_u \times \mathbf{r}_v}{|\mathbf{r}_u \times \mathbf{r}_v|} \tag{5.40}$$

is of unit magnitude and is perpendicular to both the u- and v-coordinate curves at P. It is called a *unit normal vector* to S.

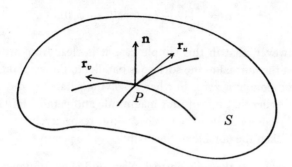

Fig. 57. The unit normal $\mathbf{n}$ to a surface S

If $\mathbf{r}_u$ and $\mathbf{r}_v$ are interchanged in (5.40), the direction of $\mathbf{n}$ is reversed, and so to this extent the direction of $\mathbf{n}$ is not unique. It is convenient to avoid this ambiguity by labelling one side of S as positive, and to take the order of $\mathbf{r}_u$ and $\mathbf{r}_v$ such that $\mathbf{n}$ points away from the positive side. *In the case of a closed surface bounding a region $\mathcal{R}$, it is conventional to label the surface so that the unit normal points out of $\mathcal{R}$* (Fig. 58).

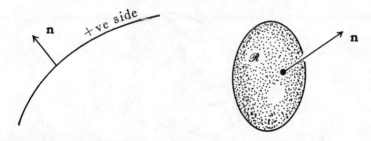

Fig. 58. Orientation of a surface. By convention the positive normal of a surface bounding a closed region $\mathcal{R}$ points out of $\mathcal{R}$

A surface is said to be *oriented* once a particular side has been labelled as positive.

To illustrate the definition (5.40), consider the circular cylinder

$$\mathbf{r} = (a\cos\phi, a\sin\phi, z) \quad (0 \leqslant \phi \leqslant 2\pi, -\infty < z < \infty).$$

In this case,

$$\mathbf{r}_\phi = (-a\sin\phi, a\cos\phi, 0)$$

and

$$\mathbf{r}_z = (0, 0, 1).$$

Thus

$$\mathbf{r}_\phi \times \mathbf{r}_z = (a \cos \phi, a \sin \phi, 0). \tag{5.41}$$

Hence,

$$\mathbf{n} = \frac{\mathbf{r}_\phi \times \mathbf{r}_z}{|\mathbf{r}_\phi \times \mathbf{r}_z|} = (\cos \phi, \sin \phi, 0). \tag{5.42}$$

Note, however, that in this simple case it is clear from a diagram (see Fig. 55) that the normal to the surface is parallel to the xy-plane, and hence that its z-component is zero. Further, resolving parallel to the axes Ox, Oy and remembering that $\mathbf{n}$ is of unit magnitude and parallel to NP, it is seen that the x-, y-components of $\mathbf{n}$ are $\cos \phi, \sin \phi$, respectively.

We have taken the outside of the cylinder to be positive.

Smooth surface. If the unit normal exists and is continuous at all points on a surface, then the surface is said to be *smooth*.

Simple surface. A *simple* surface is one which is the union of a finite number of smooth surfaces. Such surfaces are also sometimes termed *piecewise-smooth* or *regular*.

The cap of a sphere is an example of a smooth surface; and the surface of a cube is a simple, closed surface, which is the union of six smooth open surfaces.

FIG. 59. A surface element dS

Surface area. Let $P_0(u, v)$ be a point on the surface S with parametric equation $\mathbf{r} = \mathbf{r}(u, v)$. Let $P_1(u + du, v)$ be a neighbouring point on the u-coordinate curve through P_0, and let $P_2(u, v + dv)$ be a neighbouring point on the v-coordinate curve through P_0. Let the u-coordinate curve through P_2 and the v-coordinate curve through P_1 meet at P_3 (Fig. 59). The part of S which is bounded by $P_0 P_1 P_3 P_2$ is called a *surface element*. If we suppose that $P_0 P_1 P_3 P_2$ is approximately a parallelogram, as intuitive geometrical considerations suggest, then the area of the surface element is

$$dS \approx |\overrightarrow{P_0 P_1} \times \overrightarrow{P_0 P_2}|$$
$$\approx |\mathbf{r}_u \times \mathbf{r}_v| \, du \, dv, \tag{5.43}$$

using the fact that $\overrightarrow{P_0 P_1} \approx (\partial \mathbf{r}/\partial u) \, du$ and $\overrightarrow{P_0 P_2} \approx (\partial \mathbf{r}/\partial v) \, dv$. This intuitive argument leads us to *define*:

$$\text{surface area of } S = \iint |\mathbf{r}_u \times \mathbf{r}_v| \, du \, dv, \tag{5.44}$$

where the ranges of u, v are such that the whole of S is covered. We shall refer to $|\mathbf{r}_u \times \mathbf{r}_v| \, du \, dv$ as *an element of surface area dS*, although, of course, the only precisely defined quantity is the total surface area.

To illustrate this definition, consider the circular cylinder

$$\mathbf{r} = (a \cos \phi, a \sin \phi, z) \quad 0 \leqslant \phi \leqslant 2\pi, \, 0 \leqslant z \leqslant b,$$

which is of length b and radius a (Fig. 60). Its area is, of course, $2\pi ab$. Using the result (5.41),

$$|\mathbf{r}_\phi \times \mathbf{r}_z| = a,$$

and thus, by (5.44),

$$\text{total area of the cylinder} = \int_0^b \int_0^{2\pi} a \, d\phi \, dz$$
$$= 2\pi ab.$$

Note that in this case, if ϕ increases by $d\phi$, P traces out an arc of length $a \, d\phi$; and if z increases by dz, P is displaced a distance dz (Fig. 60). Since these displacements are at right angles, it is evident that the area of a surface element is

$$dS = a \, d\phi \, dz, \tag{5.45}$$

which leads to the above integral for the total surface area.

In the case of the sphere defined by (5.38), similar geometrical considerations (Fig. 61) lead to the conclusion that

$$dS = a^2 \sin \theta \, d\theta \, d\phi. \tag{5.46}$$

Thus,

$$\text{total surface area of the sphere} = \int_0^{2\pi} \int_0^{\pi} a^2 \sin \theta \, d\theta \, d\phi$$
$$= 4\pi a^2,$$

which again agrees with the familiar elementary result.

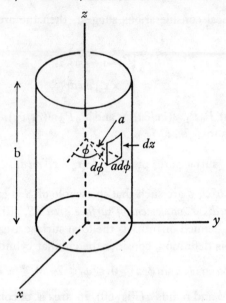

FIG. 60. Element of surface area on a cylinder

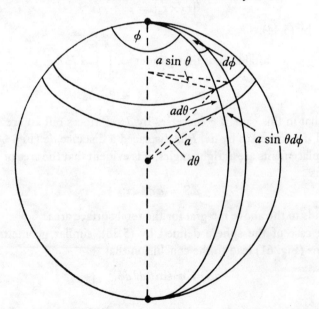

FIG. 61. Element of surface area on a sphere

Coordinate surfaces. Let u, v, w be curvilinear coordinates, defined relative to a rectangular cartesian system by the relation

$$\mathbf{r} = \mathbf{r}(u, v, w). \tag{5.47}$$

If u_0, v_0, w_0 are constants, then each of the equations

$$\mathbf{r} = \mathbf{r}(u_0, v, w), \quad \mathbf{r} = \mathbf{r}(u, v_0, w), \quad \mathbf{r} = \mathbf{r}(u, v, w_0) \tag{5.48}$$

represents a surface: these surfaces are called u-, v-, w-coordinate surfaces, respectively.

When the system of curvilinear coordinates is orthogonal, the unit normal vectors and the elements of area of the coordinate surfaces can be conveniently expressed in terms of quantities defined in § 4.11, p. 102. The unit normal to a w-coordinate surface is

$$\mathbf{n} = \frac{\mathbf{r}_u \times \mathbf{r}_v}{|\mathbf{r}_u \times \mathbf{r}_v|} = \frac{\mathbf{r}_u}{|\mathbf{r}_u|} \times \frac{\mathbf{r}_v}{|\mathbf{r}_v|},$$

where we have used the fact that $\mathbf{r}_u$ and $\mathbf{r}_v$ are perpendicular. But from the definitions (4.71), $h_1 = |\mathbf{r}_u|$ and $h_2 = |\mathbf{r}_v|$. Hence, using the definitions (4.72) and noting that $\mathbf{e}_u$, $\mathbf{e}_v$, $\mathbf{e}_w$ form a right-handed orthonormal triad,

$$\mathbf{n} = \mathbf{e}_u \times \mathbf{e}_v = \mathbf{e}_w.$$

Similarly, the unit normals to the u- and v-coordinate surfaces are $\mathbf{e}_u$ and $\mathbf{e}_v$ respectively. (The particular orientation of the coordinate surfaces which is implied here is conventional.)

The element of area of a w-coordinate surface is

$$\begin{aligned} dS &= |\mathbf{r}_u \times \mathbf{r}_v| \, du \, dv \\ &= h_1 h_2 \, du \, dv; \end{aligned}$$

and similarly the elements of area of the u-, v-coordinate surfaces are $h_2 h_3 \, dv \, dw$, $h_1 h_3 \, du \, dw$, respectively.

For example, consider cylindrical polar coordinates R, ϕ, z, defined by

$$\mathbf{r} = (R\cos\phi, R\sin\phi, z).$$

The coordinate surface $R = a$ is a circular cylinder with axis Oz. By what has been said above, and using equations (4.77) on p. 105, the unit normal to this surface is

$$\mathbf{n} = \mathbf{e}_R = (\cos\phi, \sin\phi, 0),$$

in agreement with (5.42). Further, from equations (4.76) on p. 105, $h_1 = 1$, $h_2 = R$, and so the element of area of the cylinder $R = a$ is

$$\begin{aligned} dS &= h_1 h_2 \, d\phi \, dz \\ &= a \, d\phi \, dz, \end{aligned}$$

in agreement with (5.45).

Alternative form of the equation of a surface. Consider again the surface with parametric equations

$$x = x(u, v), \quad y = y(u, v), \quad z = z(u, v).$$

If we suppose that the first two equations are solved so as to give u and v as functions of x and y, and that these functions are substituted into the third equation, we obtain

$$z = f(x, y) \tag{5.49}$$

as the equation of the surface. This representation was referred to earlier in this chapter (see p. 129). It can be shown to be a possible representation provided no portion (of non-zero area) of the surface is composed of lines parallel to the z-axis; that is, provided the normal is not parallel to the xy-plane. Clearly, similar remarks can be made about the possibilities of representing the surface in the form $x = g(y, z)$ or $y = h(z, x)$.

For example, the portion of the sphere of radius a, centre the origin, which lies in the region $z \geqslant 0$ can be represented by

$$z = (a^2 - x^2 - y^2)^{\frac{1}{2}}, \quad (x^2 + y^2 \leqslant a^2),$$

which is of the form (5.49). On the other hand, the equation of a circular cylinder of radius a and axis Oz cannot be expressed in the form (5.49). Its rectangular cartesian equation is $x^2 + y^2 = a^2$, the variable z being absent.

EXERCISES

27. Using the definition (5.40), verify that the unit normal to the sphere

$$\mathbf{r} = (a \sin \theta \cos \phi, a \sin \theta \sin \phi, a \cos \theta)$$

is

$$\mathbf{n} = (\sin \theta \cos \phi, \sin \theta \sin \phi, \cos \theta).$$

Also, using the definition (5.44), verify that the surface area of the lune of the sphere between the planes $\phi = \phi_0$, $\phi = \phi_0 + \alpha$ is $2\alpha a^2$.

28. Show that the surface

$$\mathbf{r} = (r \cos \phi, r \sin \phi, r\sqrt{3}),$$

where $0 \leqslant r < \infty$, $0 \leqslant \phi \leqslant 2\pi$, is an infinite cone with vertex at the origin, with axis along Oz, and having a semi-vertical angle of $30°$. Find (i) the unit normal vector $\mathbf{n}$ at any point, and (ii) the element of surface area.

29. Describe (in geometrical terms) the surfaces whose parametric equations are:

(i) $\mathbf{r} = (a \cos u, b \sin u, v), \quad 0 \leqslant u \leqslant 2\pi, \quad -\infty < v < \infty;$

(ii) $\mathbf{r} = (a \cosh u, b \sinh u, v), \quad -\infty < u < \infty \quad -\infty < v < \infty.$

In case (i), find, in the form of an integral with respect to u, the total surface area between the planes $z = 0$, $z = 2$.

30. Find the total surface area of the surface with parametric equation

$$\mathbf{r} = (u \cos v, u \sin v, u^2), \quad 0 \leqslant u \leqslant 1, \quad 0 \leqslant v \leqslant 2\pi.$$

31. Express the equation of the conical surface defined in Exercise 28 in the form $z = f(x, y)$.

32. The projection on the xy-plane of the surface $z = f(x, y)$ is the region $\mathcal{R}$. Show that the area of the surface is given by

$$\iint_{\mathcal{R}} (1 + f_x^2 + f_y^2)^{\frac{1}{2}} \, dx \, dy,$$

where suffixes denote partial derivatives with respect to x, y. Also, show that the unit normal at any point is

$$\pm \frac{(f_x, f_y, -1)}{\sqrt{(f_x^2 + f_y^2 + 1)}}.$$

[*Hint.* The parametric equation of the surface may be taken as $\mathbf{r} = (x, y, f(x, y))$.]

5.6 Surface integrals

Let S be a simple surface with parametric equation $\mathbf{r} = \mathbf{r}(u, v)$, and let $\mathcal{R}$ denote the region in the uv-plane consisting of points corresponding to the points of S. If a scalar field Ω and a vector field $\mathbf{F}$ are defined at all points on S, then, on S, $\Omega = \Omega(u, v)$ and $\mathbf{F} = \mathbf{F}(u, v)$.

We define the surface integrals of Ω and $\mathbf{F}$ over S as follows:

$$\int_S \Omega \, dS = \iint_{\mathcal{R}} \Omega(u, v) \, |\mathbf{r}_u \times \mathbf{r}_v| \, du \, dv; \tag{5.50}$$

$$\int_S \mathbf{F} . \, d\mathbf{S} = \int_S \mathbf{F} . \mathbf{n} \, dS = \iint_{\mathcal{R}} \mathbf{F}(u, v) . (\mathbf{r}_u \times \mathbf{r}_v) \, du \, dv. \tag{5.51}$$

Thus *dS is interpreted as an element of surface area*; *and* $\mathbf{dS}$ *is an abbreviation for* $\mathbf{n} \, dS$, *where* $\mathbf{n} = (\mathbf{r}_u \times \mathbf{r}_v) / |\mathbf{r}_u \times \mathbf{r}_v|$ *is the unit normal vector.*

If $\mathbf{F} = F_1 \mathbf{i} + F_2 \mathbf{j} + F_3 \mathbf{k}$, in a rectangular cartesian coordinate system, we also define

$$\int_S \mathbf{F} \, dS = \left(\int_S F_1 \, dS \right) \mathbf{i} + \left(\int_S F_2 \, dS \right) \mathbf{j} + \left(\int_S F_3 \, dS \right) \mathbf{k}. \tag{5.52}$$

The unit normal vector $\mathbf{n}$ is itself a vector field defined on S. Thus the integrals

$$\int_S \Omega \, \mathbf{dS} = \int_S \Omega \mathbf{n} \, dS \tag{5.53}$$

and

$$\int_S \mathbf{F} \times \mathbf{dS} = \int_S \mathbf{F} \times \mathbf{n} \, dS \tag{5.54}$$

are also now meaningful in view of the definition (5.52).

Note. Many authors use the slightly modified notation

$$\iint_S \Omega \, dS$$

to denote a surface integral, indicating that it is really a double integral. Likewise, in the case of volume integrals, defined in the next section, three integral signs are often used to indicate that a volume integral is a triple integral. However, the shorter notation used here can lead to no confusion if the definitions of surface and volume integrals are borne in mind.

EXAMPLE 15. Evaluate

$$\int_S \Omega \, dS$$

in the following cases:

(i) $\Omega = x^2 + y^2$ and S is the spherical surface $x^2 + y^2 + z^2 = a^2$;
(ii) $\Omega = x^2 + y^2$ and S is the complete surface of the cube $|x| \leqslant a, |y| \leqslant a, |z| \leqslant a$;
(iii) $\Omega = (b^2 x^2 / a^2 + a^2 y^2 / b^2)^{\frac{1}{2}}$ and S is the curved surface of the elliptic cylinder $x^2 / a^2 + y^2 / b^2 = 1, |z| \leqslant c$.

Solution. (i) The parametric equations of the sphere S are

$$x = a \sin \theta \cos \phi, \quad y = a \sin \theta \sin \phi, \quad z = a \cos \theta,$$

where $0 \leqslant \theta \leqslant \pi, 0 \leqslant \phi \leqslant 2\pi$. Thus, on S

$$\Omega = a^2 \sin^2 \theta.$$

We also have $dS = a^2 \sin \theta \, d\theta \, d\phi$. Hence

$$\int_S \Omega \, dS = \int_0^{2\pi} \int_0^{\pi} a^4 \sin^3 \theta \, d\theta \, d\phi$$

$$= \tfrac{8}{3} \pi a^4.$$

(ii) This integral is evaluated by considering the six separate faces of the cube. On the two faces $z = \pm a$, parallel to the xy-plane, the element of surface area is $dS = dx \, dy$. Thus, the joint contribution to the integral from this pair of faces is

$$C_1 = 2 \int_{-a}^{a} \int_{-a}^{a} (x^2 + y^2) \, dx \, dy$$

$$= \tfrac{16}{3} a^4.$$

On the faces $x = \pm a$, $\Omega = a^2 + y^2$ and $dS = dy \, dz$. Thus the joint contribution from this pair of faces is

$$C_2 = 2 \int_{-a}^{a} \int_{-a}^{a} (a^2 + y^2) \, dy \, dz$$

$$= \tfrac{32}{3} a^4.$$

Finally, the joint contribution from the pair of faces $y = \pm a$ is

$$C_3 = 2 \int\limits_{-a}^{a} \int\limits_{-a}^{a} (x^2 + a^2)\, dx\, dz$$

$$= \tfrac{3\,2}{3} a^4.$$

Hence,

$$\int\limits_{S} \Omega\, dS = C_1 + C_2 + C_3 = \tfrac{8\,0}{3} a^4.$$

(iii) The parametric equation of the elliptic cylindrical surface can be taken as

$$\mathbf{r} = (a\cos\theta, b\sin\theta, z)$$

where $0 \leqslant \theta \leqslant 2\pi$, $-c \leqslant z \leqslant c$. Then

$$\mathbf{r}_\theta = (-a\sin\theta, b\cos\theta, 0),$$
$$\mathbf{r}_z = (0, 0, 1),$$

giving

$$|\mathbf{r}_\theta \times \mathbf{r}_z| = |(b\cos\theta, a\sin\theta, 0)|$$
$$= (b^2\cos^2\theta + a^2\sin^2\theta)^{\frac{1}{2}}.$$

Thus

$$dS = (b^2\cos^2\theta + a^2\sin^2\theta)^{\frac{1}{2}}\, d\theta\, dz.$$

On S,

$$\Omega = (b^2\cos^2\theta + a^2\sin^2\theta)^{\frac{1}{2}},$$

and hence

$$\int\limits_{S} \Omega\, dS = \int\limits_{-c}^{c} \int\limits_{0}^{2\pi} (b^2\cos^2\theta + a^2\sin^2\theta)\, d\theta\, dz$$

$$= 2\pi c(a^2 + b^2).$$

EXAMPLE 16. Let $\mathbf{r}$ denote the position vector of a point P relative to the origin O. Evaluate

$$\int\limits_{S} \mathbf{r} . d\mathbf{S},$$

where S is the curved surface of the cylinder $x^2 + y^2 = a^2$, $0 \leqslant z \leqslant 2a$.

Solution. By simple geometrical considerations (Fig. 62), when P lies on S,

$$\mathbf{r} . \mathbf{n} = OP\cos\alpha \quad (\alpha = \text{angle between } \mathbf{r} \text{ and the unit vector } \mathbf{n})$$
$$= a.$$

Thus

$$\int\limits_{S} \mathbf{r} . d\mathbf{S} = \int\limits_{S} a\, dS$$

$$= a \times (\text{area of } S)$$

$$= 4\pi a^3.$$

EXAMPLE 17. If S denotes the surface of the sphere

$$\mathbf{r} = (a\sin\theta\cos\phi, a\sin\theta\sin\phi, a\cos\theta),$$

where $0 \leqslant \theta \leqslant \pi$, $0 \leqslant \phi \leqslant 2\pi$, show that

$$\int_S (x^2+y^2)\,\mathbf{dS} = \mathbf{0}.$$

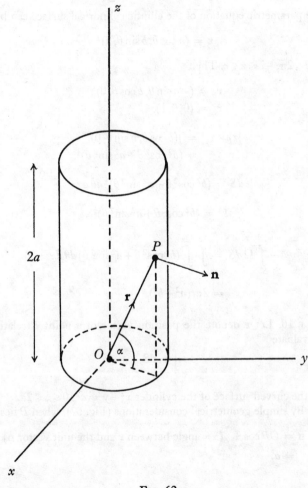

FIG. 62

Solution. The unit normal to the sphere will be directed away from its centre, which is the origin O. Thus

$$\mathbf{n} = \hat{\mathbf{r}} = \mathbf{i}\sin\theta\cos\phi + \mathbf{i}\sin\theta\sin\phi + \mathbf{k}\cos\theta.$$

(This result can of course be deduced from the definition (5.40) of the unit normal.) Also $dS = a^2 \sin\theta\, d\theta\, d\phi$, and on S, $x^2 + y^2 = a^2 \sin^2\theta$. Thus

$$\int_S (x^2+y^2)\, \mathbf{dS} = \int_0^{2\pi}\int_0^{\pi} a^4 \sin^3\theta\, (\mathbf{i}\sin\theta\cos\phi + \mathbf{j}\sin\theta\sin\phi + \mathbf{k}\cos\theta)\, d\theta\, d\phi$$

$$= \mathbf{i}\int_0^{2\pi}\int_0^{\pi} a^4 \sin^4\theta\cos\phi\, d\theta\, d\phi$$

$$+ \mathbf{j}\int_0^{2\pi}\int_0^{\pi} a^4 \sin^4\theta\sin\phi\, d\theta\, d\phi$$

$$+ \mathbf{k}\int_0^{2\pi}\int_0^{\pi} a^4 \sin^3\theta\cos\theta\, d\theta\, d\phi.$$

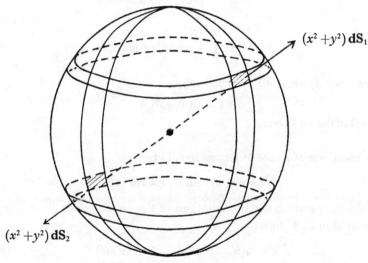

$(x^2 + y^2)\, \mathbf{dS}_1$

$(x^2 + y^2)\, \mathbf{dS}_2$

FIG. 63

Integrating with respect to ϕ first, and observing that

$$\int_0^{2\pi} \cos\phi\, d\phi = \int_0^{2\pi} \sin\phi\, d\phi = 0$$

gives

$$\int_S (x^2+y^2)\, \mathbf{dS} = 2\pi a^4 [\tfrac{1}{4}\sin^4\theta]_0^{\pi}\, \mathbf{k}$$

$$= \mathbf{0}.$$

Note. The above result can be deduced by a simple symmetry argument; for Fig. 63 shows that the contributions to the integral from surface elements $\mathbf{dS}_1$ and $\mathbf{dS}_2$ at opposite ends of a diameter cancel. The value of $x^2 + y^2$ is the same at each end of the diameter.

II

EXAMPLE 18. Evaluate

$$\mathbf{I} = \int_S \mathbf{r} \, dS,$$

where S denotes the hemisphere $x^2 + y^2 + z^2 = a^2$, $z \geqslant 0$, and $\mathbf{r}$ is the radius vector from the origin.

Solution. The parametric equation of S is

$$\mathbf{r} = (a \sin \theta \cos \phi, a \sin \theta \sin \phi, a \cos \theta),$$

where $0 \leqslant \theta \leqslant \frac{1}{2}\pi$, $0 \leqslant \phi \leqslant 2\pi$. Since $dS = a^2 \sin \theta \, d\theta \, d\phi$, the three components of $\mathbf{I}$ are thus

$$I_1 = a^3 \int_0^{2\pi} \int_0^{\frac{1}{2}\pi} \sin^2 \theta \cos \phi \, d\theta \, d\phi,$$

$$I_2 = a^3 \int_0^{2\pi} \int_0^{\frac{1}{2}\pi} \sin^2 \theta \sin \phi \, d\theta \, d\phi,$$

$$I_3 = a^3 \int_0^{2\pi} \int_0^{\frac{1}{2}\pi} \sin \theta \cos \theta \, d\theta \, d\phi.$$

Hence $I_1 = 0$, $I_2 = 0$, and $I_3 = \pi a^3$, giving

$$\mathbf{I} = (0, 0, \pi a^3),$$

referred to the axes $Oxyz$.

Independence of choice of parameters. Let

$$\mathbf{r} = \mathbf{r}(u, v) \quad \text{and} \quad \mathbf{r} = \mathbf{r}(u', v') \tag{5.55}$$

be different parametric representations of the same surface S, so that u, v are related to u', v' thus:

$$u = u(u', v') \quad \text{and} \quad v = v(u', v'). \tag{5.56}$$

Then

$$\begin{aligned}
\mathbf{r}_{u'} \times \mathbf{r}_{v'} &= (x_{u'}, y_{u'}, z_{u'}) \times (x_{v'}, y_{v'}, z_{v'}) \\
&= (y_{u'} z_{v'} - y_{v'} z_{u'}, z_{u'} x_{v'} - z_{v'} x_{u'}, x_{u'} y_{v'} - x_{v'} y_{u'}) \\
&= \left(\frac{\partial(y, z)}{\partial(u', v')}, \frac{\partial(z, x)}{\partial(u', v')}, \frac{\partial(x, y)}{\partial(u', v')} \right),
\end{aligned}$$

using the Jacobian notation introduced on p. 131 (see equation (5.30)). By the chain rule for Jacobians, proved in Appendix 2, it follows that

$$\begin{aligned}
\mathbf{r}_{u'} \times \mathbf{r}_{v'} &= \left(\frac{\partial(y, z)}{\partial(u, v)}, \frac{\partial(z, x)}{\partial(u, v)}, \frac{\partial(x, y)}{\partial(u, v)} \right) \frac{\partial(u, v)}{\partial(u', v')} \\
&= (\mathbf{r}_u \times \mathbf{r}_v) J,
\end{aligned}$$

where J denotes the Jacobian of the transformation (5.56). Using this result and the formula (5.29) for changing the variables in a double integral, we thus find

$$\iint\limits_{\mathscr{R}} \phi(u, v)|\mathbf{r}_u \times \mathbf{r}_v|\, du\, dv = \iint\limits_{\mathscr{R}'} \phi(u', v')|\mathbf{r}_{u'} \times \mathbf{r}_{v'}|\, du'\, dv',$$

where the regions of integration $\mathscr{R}$, $\mathscr{R}'$, each correspond to S. But, by definition, these are the surface integrals of ϕ over S in terms of the two parametric representations (5.55). It follows that *the integral of ϕ over S is independent of the particular choice of parameters on S*. The significance of this result is perhaps made clearer by considering a simple physical situation.

Let P denote the *downward component* of force per unit area acting on a surface S. On an element of surface area dS, the downward force will be $P\, dS$ and hence the total downward force on S will be

$$\int\limits_{S} P\, dS.$$

To evaluate this integral it would be necessary to choose a particular parametric representation of S, but on physical grounds the final result should be independent of the choice made. The result proved above confirms this.

A surface integral involving the vector element of area $\mathbf{dS} = \mathbf{n}\, dS$ is again independent of the choice of parameters, but if the orientation is changed the direction of $\mathbf{n}$ is reversed, and the integral changes sign.

EXERCISES

33. Evaluate

$$\int\limits_{S} xyz\, dS$$

over that part of a sphere of unit radius and centre at the origin which lies in the positive octant $x \geqslant 0$, $y \geqslant 0$, $z \geqslant 0$.

34. The ends of a circular cylinder of radius a and axis Oz are in the planes $z = 0$, $z = 2a$. If S denotes the complete surface of the cylinder (including the plane ends), evaluate

$$\int\limits_{S} z\, dS.$$

35. If S denotes the surface of the hemisphere $x^2 + y^2 + z^2 = 1$, $z \geqslant 0$, find

$$\text{(i)} \int\limits_{S} \mathbf{F} \cdot \mathbf{dS}, \qquad \text{(ii)} \int\limits_{S} (\mathbf{r} \times \mathbf{k}) \times \mathbf{dS},$$

where
$$\mathbf{F} = (x^2, y^2, z^2), \quad \mathbf{r} = (x, y, z), \quad \mathbf{k} = (0, 0, 1),$$

and the unit normal to S points away from the origin.

36. Evaluate

$$\text{(i)} \int_S \mathbf{F}\, dS, \quad \text{(ii)} \int_S \mathbf{F}.\, d\mathbf{S}, \quad \text{(iii)} \int_S \mathbf{F} \times d\mathbf{S},$$

where
$$\mathbf{F} = (y+z, z+x, x+y),$$
and S is the square
$$0 \leqslant x \leqslant 1, \quad 0 \leqslant y \leqslant 1, \quad z = 0,$$

positively oriented along the positive z-axis.

37. If S denotes that part of the conical surface

$$\mathbf{r} = (r \cos \phi, r \sin \phi, r\sqrt{3})$$

lying between the planes $z = 0$, $z = 1$, evaluate

$$\int_S (1 - z)\, d\mathbf{S}.$$

[*Hint.* Cf. Exercise 28, p. 146.]

38. If S denotes that part of the plane $2x + y + 2z = 6$ which lies in the positive octant, and $\mathbf{F} = (4x, y, z)$, evaluate

$$\int_S \mathbf{F}.\, d\mathbf{S}.$$

Take the unit normal to S pointing away from the origin. [*Hint.* The parametric equation of the plane can be taken as

$$\mathbf{r} = (x, y, 3 - x - \tfrac{1}{2}y).]$$

5.7 Volume integrals

Let u, v, w be a system of curvilinear coordinates such that, relative to a rectangular cartesian coordinate system, the position vector of a point is

$$\mathbf{r} = \mathbf{r}(u, v, w). \tag{5.57}$$

Let τ be a region in the rectangular cartesian space, and let $\mathscr{R}$ be the region in the *uvw*-space consisting of points corresponding to the points of τ. Then the volume V of τ is defined as

$$V = \iiint_{\mathscr{R}} |(\mathbf{r}_u \times \mathbf{r}_v).\mathbf{r}_w|\, du\, dv\, dw. \tag{5.58}$$

Further, if Ω is a scalar field defined on τ, the volume integral of Ω over τ is defined as

$$\int_\tau \Omega \, d\tau = \iiint_\mathscr{R} \Omega(u, v, w) |(\mathbf{r}_u \times \mathbf{r}_v) . \mathbf{r}_w| \, du \, dv \, dw. \tag{5.59}$$

It can be verified, by using the transformation formula (5.34) for triple integrals, that the integrals (5.58), (5.59) are invariant under a change of curvilinear coordinates (cf. the corresponding result for surface integrals, proved in the previous section).

The motivation for the above definition of volume is as follows.

If u is given a small (positive) increment du, the point $P_0(u, v, w)$ is displaced along a u-coordinate curve to $P_1(u+du, v, w)$, and $\overrightarrow{P_0 P_1} \approx \mathbf{r}_u du$.

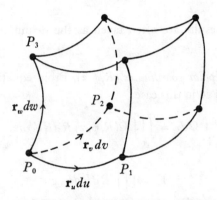

FIG. 64. An element of volume; $\overrightarrow{P_0 P_1} \approx \mathbf{r}_u du$, etc.

Similarly, corresponding to increments dv in v and dw in w, P is displaced to P_2 and P_3 respectively, where $\overrightarrow{P_0 P_2} \approx \mathbf{r}_v dv$ and $\overrightarrow{P_0 P_3} \approx \mathbf{r}_w dw$ (Fig. 64). The element bounded by the various coordinate surfaces through P_0, P_1, P_2, P_3 is approximately a parallelepiped of volume

$$\begin{aligned} d\tau &= |(\overrightarrow{P_0 P_1} \times \overrightarrow{P_0 P_2}) . \overrightarrow{P_0 P_3}| \\ &= |(\mathbf{r}_u \times \mathbf{r}_v) . \mathbf{r}_w| \, du \, dv \, dw. \end{aligned} \tag{5.60}$$

As the triple integral (5.58) is (by definition) the limit as $du \to 0$, $dv \to 0$, $dw \to 0$ of a sum of terms of the form (5.60), the definition of volume agrees with the intuitive notion.

When u, v, w are orthogonal curvilinear coordinates,

$$(\mathbf{r}_u \times \mathbf{r}_v) . \mathbf{r}_w = h_1 h_2 h_3 (\mathbf{e}_u \times \mathbf{e}_v) . \mathbf{e}_w = h_1 h_2 h_3,$$

in the usual notation p.103, and so

$$\int_\tau \Omega \, d\tau = \iiint_{\mathscr{R}} \Omega(u, v, w) \, h_1 h_2 h_3 \, du \, dv \, dw. \tag{5.61}$$

There are three important particular cases.

(i) *Rectangular cartesian coordinates* (x,y,z). In this case $h_1 = h_2 = h_3 = 1$, and hence

$$\int_\tau \Omega \, d\tau = \iiint_\tau \Omega(x, y, z) \, dx \, dy \, dz. \tag{5.62}$$

Also, the volume of τ is simply

$$V = \iiint_\tau dx \, dy \, dz. \tag{5.63}$$

Note that these results are often taken as the definitions of volume and volume integral.

(ii) *Cylindrical polar coordinates* (R, ϕ, z). From equations (4.76) $h_1 = 1$, $h_2 = R$, $h_3 = 1$, and so in this case

$$\int_\tau \Omega \, d\tau = \iiint_{\mathscr{R}} \Omega(R, \phi, z) \, R \, dR \, d\phi \, dz. \tag{5.64}$$

Also

$$V = \iiint_{\mathscr{R}} R \, dR \, d\phi \, dz \tag{5.65}$$

gives the volume of τ.

(iii) *Spherical polar coordinates* (r, θ, ϕ). In this case equations (4.74) give $h_1 = 1$, $h_2 = r$, $h_3 = r \sin \theta$, and hence

$$\int_\tau \Omega \, d\tau = \iiint_{\mathscr{R}} \Omega(r, \theta, \phi) \, r^2 \sin \theta \, dr \, d\theta \, d\phi. \tag{5.66}$$

Also

$$V = \iiint_{\mathscr{R}} r^2 \sin \theta \, dr \, d\theta \, d\phi. \tag{5.67}$$

EXAMPLE 19. Evaluate

$$I = \int_\tau \exp[-(x^2+y^2+z^2)] \, d\tau,$$

where τ denotes the whole of space.

Solution. To cover the whole of space, let each of x, y, z range from $-\infty$ to ∞. Then

$$I = \int\limits_{-\infty}^{\infty} \int\limits_{-\infty}^{\infty} \int\limits_{-\infty}^{\infty} \exp[-(x^2+y^2+z^2)]\,dx\,dy\,dz$$

$$= \left(\int\limits_{-\infty}^{\infty} e^{-x^2}\,dx\right)\left(\int\limits_{-\infty}^{\infty} e^{-y^2}\,dy\right)\left(\int\limits_{-\infty}^{\infty} e^{-z^2}\,dz\right).$$

But, using a result proved in Example 9, p. 131,

$$\int\limits_{-\infty}^{\infty} e^{-x^2}\,dx = 2\int\limits_{0}^{\infty} e^{-x^2}\,dx = \sqrt{\pi}.$$

Thus,

$$I = \pi^{3/2}.$$

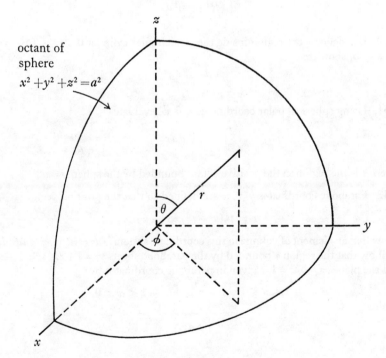

octant of
sphere
$x^2+y^2+z^2=a^2$

FIG. 65

EXAMPLE 20. Let τ denote the octant of a sphere defined by $x^2+y^2+z^2 \leqslant a^2$, $x\geqslant 0$, $y\geqslant 0$, $z\geqslant 0$. Evaluate

$$I = \int\limits_{\tau} r^2\,d\tau.$$

Solution. The region of integration τ is shown in Fig. 65, and is covered by letting r range from 0 to a, θ range from 0 to $\frac{1}{2}\pi$, and ϕ range from 0 to $\frac{1}{2}\pi$. Thus, as $d\tau = r^2 \sin\theta\, dr\, d\theta\, d\phi$,

$$I = \int_0^{\frac{1}{2}\pi} \int_0^{\frac{1}{2}\pi} \int_0^a r^4 \sin\theta\, dr\, d\theta\, d\phi$$

$$= \left(\int_0^a r^4\, dr \right) \left(\int_0^{\frac{1}{2}\pi} \sin\theta\, d\theta \right) \left(\int_0^{\frac{1}{2}\pi} d\phi \right)$$

$$= \tfrac{1}{10}\pi a^5.$$

EXERCISES

39. If τ denotes the region $R \leqslant a$, $0 \leqslant z \leqslant b$ in cylindrical polar coordinates R, ϕ, z, evaluate

$$\int_\tau (R^2 + bz)^{\frac{1}{2}}\, d\tau.$$

40. If τ denotes the region inside the semicircular cylinder $0 \leqslant x \leqslant \sqrt{(a^2 - y^2)}$, $0 \leqslant z \leqslant 2a$, show that

$$\int_\tau x\, d\tau = \tfrac{4}{3}a^4.$$

41. Using spherical polar coordinates r, θ, ϕ, evaluate

$$\int_\tau r^2 \sin^2\theta\, d\tau,$$

where τ is the region in the positive octant bounded by the sphere $r = a$.

42. Parabolic coordinates u, v, w are defined so that the position vector is

$$\mathbf{r} = (\tfrac{1}{2}(u^2 - v^2), uv, -w).$$

Show that an element of volume in this coordinate system is $d\tau = (u^2 + v^2)\, du\, dv\, dw$.

Show that the region τ bounded by the parabolic sheets $y^2 = 1 + 2x$, $y^2 = 1 - 2x$, and the planes $z = 0$, $z = 1$ is given in parabolic coordinates by

$$-1 \leqslant u \leqslant 1, \quad 0 \leqslant v \leqslant 1, \quad -1 \leqslant w \leqslant 0.$$

Evaluate

$$\int_\tau z\, d\tau.$$

CHAPTER 6

INTEGRAL THEOREMS

6.1 Introduction

In Chapter 4, scalar and vector fields were defined and the properties of gradient, divergence, and curl were discussed; and in Chapter 5 we were concerned with various integrals of scalar and vector fields, and the techniques whereby such integrals are evaluated. The ground has thus been prepared for the two *central theorems in vector analysis*: (i) *The divergence theorem* (also called Gauss's theorem), which relates the integral of a vector field **F** over a closed surface S to the volume integral of div**F** over the region bounded by S; and (ii) *Stokes's theorem* which relates the integral of a vector field **F** around a closed curve $\mathscr{C}$ to the integral of curl**F** over any open surface S bounded by $\mathscr{C}$. In this chapter we shall prove these theorems and some related results.

6.2 The Divergence Theorem (Gauss's theorem)

Definitions. 1. A simple closed surface S is said to be *convex* if any straight line meeting it does so in at most two points.

2. A simple closed surface S will be called a *semi-convex surface* if axes $Oxyz$ can be so chosen that any line drawn parallel to one of the axes and meeting S either (i) does so in just one or just two points, or (ii) has a portion of finite length in common with S.

The divergence theorem. A closed region τ is bounded by a simple closed surface S. If the vector field **F** and its divergence are defined throughout τ, then

$$\int_S \mathbf{F} \cdot d\mathbf{S} = \int_\tau \operatorname{div} \mathbf{F} \, d\tau. \tag{6.1}$$

Proof. The result will first be proved, in part (a), for a region τ bounded by a convex surface S. The theorem will then be extended to more general regions in parts (b), (c) and (d).

(a) Take rectangular axes $Oxyz$ and denote by $\mathscr{R}$ the projection of S onto the xy-plane. Divide S into upper and lower parts, S_U and S_L, as viewed from $\mathscr{R}$ (Fig. 66). Let a line drawn parallel to Oz from a point in $\mathscr{R}$ cut S_L at P

and S_U at Q, and denote the z-coordinates of P, Q by z_P, z_Q ($z_P < z_Q$). All points (x, y, z) in τ are covered if x and y range over all values in $\mathscr{R}$ and the corresponding values of z range from z_P to z_Q.

Denoting the components of $\mathbf{F}$ relative to the axes $Oxyz$ by F_1, F_2, F_3, we have

$$\int_\tau \frac{\partial F_3}{\partial z}\, d\tau = \iiint_\tau \frac{\partial F_3}{\partial z}\, dx\, dy\, dz.$$

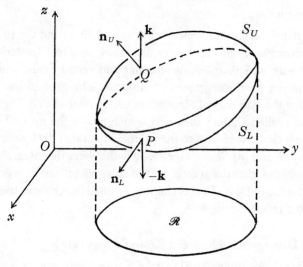

FIG. 66

Performing the integration with respect to z, this becomes

$$\int_\tau \frac{\partial F_3}{\partial z}\, d\tau = \iint_{\mathscr{R}} [F_3(x, y, z)]_{z=z_P}^{z=z_Q}\, dx\, dy$$

$$= \iint_{\mathscr{R}} \{F_3(x, y, z_Q) - F_3(x, y, z_P)\}\, dx\, dy. \qquad (6.2)$$

The integral on the right-hand side of (6.2) will now be expressed as an integral over the surface S.

Suppose that the equation of S_U is $z = f(x, y)$ (cf. p. 146). Then a particular parametric representation of S_U is

$$\mathbf{r} = [x, y, f(x, y)],$$

where x, y range over values in $\mathscr{R}$. The unit normal to S_U is therefore

$$\begin{aligned}
\mathbf{n}_U &= \pm(\mathbf{r}_x \times \mathbf{r}_y)/|\mathbf{r}_x \times \mathbf{r}_y| \\
&= \pm(1, 0, f_x) \times (0, 1, f_y)/|\mathbf{r}_x \times \mathbf{r}_y| \\
&= \pm(-f_x, -f_y, 1)/(f_x^2 + f_y^2 + 1)^{\frac{1}{2}},
\end{aligned}$$

where suffixes refer to partial derivatives with respect to x, y. As we noted on p. 141, it is conventional to choose the sign of the unit normal to a closed surface such as S so that it points away from the region enclosed. The z-component of the unit normal to the part S_U must therefore be positive, and so

$$\mathbf{n}_U = (-f_x, -f_y, 1)/(f_x^2 + f_y^2 + 1)^{\frac{1}{2}}.$$

Denoting the element of area of S_U by dS_U, it follows that

$$\mathbf{dS}_U = \mathbf{n}_U \, dS_U = \mathbf{r}_x \times \mathbf{r}_y \, dx \, dy$$
$$= (-f_x, -f_y, 1) \, dx \, dy.$$

Hence

$$\mathbf{k} \cdot \mathbf{dS}_U = dx \, dy,$$

where, as usual, $\mathbf{k} = (0, 0, 1)$ is the unit vector parallel to the z-axis. Noting that the z-component of the unit normal to the lower part S_L will be negative, it also follows by similar reasoning that

$$\mathbf{k} \cdot \mathbf{dS}_L = -dx \, dy.$$

Substituting these expressions into (6.2) and then noting that $S_U \cup S_L = S$ (see footnote), we thus have

$$\int_\tau \frac{\partial F_3}{\partial z} \, d\tau = \int_{S_U} F_3 \mathbf{k} \cdot \mathbf{dS}_U + \int_{S_L} F_3 \mathbf{k} \cdot \mathbf{dS}_L$$

$$= \int_S F_3 \mathbf{k} \cdot \mathbf{dS}. \tag{6.3}$$

Similarly,

$$\int_\tau \frac{\partial F_2}{\partial y} \, d\tau = \int_S F_2 \mathbf{j} \cdot \mathbf{dS} \tag{6.4}$$

and

$$\int_\tau \frac{\partial F_1}{\partial x} \, d\tau = \int_S F_1 \mathbf{i} \cdot \mathbf{dS}. \tag{6.5}$$

By adding (6.3), (6.4) and (6.5) we obtain

$$\int_\tau \left(\frac{\partial F_1}{\partial x} + \frac{\partial F_2}{\partial y} + \frac{\partial F_3}{\partial z} \right) d\tau = \int_S (F_1 \mathbf{i} + F_2 \mathbf{j} + F_3 \mathbf{k}) \cdot \mathbf{dS};$$

that is

$$\int_\tau \operatorname{div} \mathbf{F} \, d\tau = \int_S \mathbf{F} \cdot \mathbf{dS},$$

which is the result required.

The expression $S_U \cup S_L = S$ means that S is 'the union of' S_U and S_L; or, in other words, that S consists of the two parts S_U and S_L.

(b) Consider next a region τ bounded by a semi-convex surface S; then parts of S may consist of lines parallel to one of the axes. It is sufficient to consider how the proof in part (a) is modified when a portion S' of S consists of lines parallel to the z-axis; we denote the remainder of S by S'' (Fig. 67).

Following the lines of the argument in part (a), we obtain instead of (6.3),

$$\int_{S''} F_3 \mathbf{k}.\mathbf{dS}'' = \int_{\tau} \frac{\partial F_3}{\partial z}\, d\tau. \qquad (6.6)$$

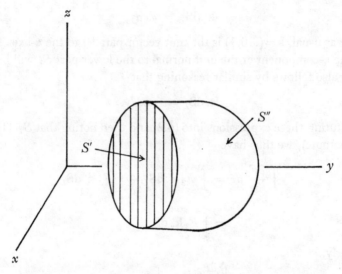

Fig. 67. S' here is composed of lines parallel to the z-axis

But on S', the unit vector $\mathbf{k}$ is everywhere perpendicular to the vector element of area $\mathbf{dS}'$, and hence

$$\int_{S'} F_3 \mathbf{k}.\mathbf{dS}' = 0. \qquad (6.7)$$

Since $S' \cup S'' = S$, we obtain (6.3) by adding (6.6) and (6.7). The remainder of the proof requires no modification, and the theorem follows as before.

(c) If the region τ does not fall into one of the classes considered in parts (a) and (b), but can be divided into regions τ_1, τ_2 which do, we have the situation shown in Fig. 68. The theorem still holds, and may be established as follows.

Let S_1, S_2 denote the bounding surfaces of τ_1, τ_2 respectively. By what we have proved so far, the divergence theorem holds for τ_1 and τ_2 considered separately. Thus,

$$\int_{S_1} \mathbf{F} . \mathbf{dS}_1 = \int_{\tau_1} \operatorname{div} \mathbf{F} \, d\tau_1$$

and

$$\int_{S_2} \mathbf{F} . \mathbf{dS}_2 = \int_{\tau_2} \operatorname{div} \mathbf{F} \, d\tau_2.$$

By addition, as $\tau_1 \cup \tau_2 = \tau$,

$$\int_{S_1} \mathbf{F} . \mathbf{dS}_1 + \int_{S_2} \mathbf{F} . \mathbf{dS}_2 = \int_{\tau} \operatorname{div} \mathbf{F} \, d\tau. \tag{6.8}$$

Denote by S_1', S_2' the parts of S_1, S_2 which also form part of S (the bounding surface of τ); and denote the remaining parts of S_1, S_2 by S_1'', S_2''

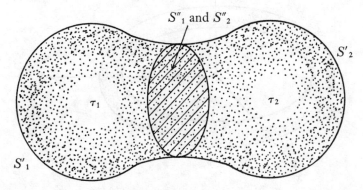

S''_1 and S''_2

S'_2

τ_1

τ_2

S'_1

FIG. 68. Here S_1'' and S_2'' are opposite sides of the shaded surface separating τ_1 from τ_2

(Fig. 68). These latter parts actually coincide, and at each point their unit normals (which are directed away from τ_1, τ_2, respectively) point in opposite directions. Thus

$$\mathbf{F} . \mathbf{dS}_1'' = -\mathbf{F} . \mathbf{dS}_2'' \tag{6.9}$$

at all points on S_1'', S_2''.

Since $S_1 = S_1' \cup S_1''$ and $S_2 = S_2' \cup S_2''$, the left-hand side of (6.8) can be expressed as

$$\int_{S_1'} \mathbf{F} . \mathbf{dS}_1' + \int_{S_1''} \mathbf{F} . \mathbf{dS}_1'' + \int_{S_2'} \mathbf{F} . \mathbf{dS}_2' + \int_{S_2''} \mathbf{F} . \mathbf{dS}_2''.$$

Using (6.9), the second and fourth integrals are seen to cancel each other, and as $S_1' \cup S_2' = S$, the sum of the remaining two integrals is

$$\int_{S} \mathbf{F} . \mathbf{dS}.$$

Thus (6.8) becomes

$$\int_S \mathbf{F}.\mathbf{dS} = \int_\tau \operatorname{div}\mathbf{F}\,d\tau,$$

which proves the theorem for this case.

(d) The most general case that we contemplate here is that of a region τ which can be subdivided into a finite number of regions τ_i $(i=1,2,\ldots,n)$ bounded by semi-convex surfaces S_i. The proof given above for $n=2$ can be extended to cover the general case; we leave the details to the reader as an exercise.

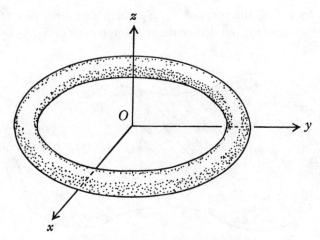

FIG. 69. A torus, or anchor ring

Extension to functions with discontinuous derivatives. In the statement of Gauss's theorem, the condition imposed upon the vector field $\mathbf{F}$ was that $\operatorname{div}\mathbf{F}$ should exist throughout the closed region τ, and this implies that $\mathbf{F}$ should be continuously differentiable in τ (cf. p. 85). However, the theorem holds under less restrictive conditions.

A particularly important case is that of a vector field $\mathbf{F}$ which is continuous in τ, but whose partial derivatives have finite discontinuities on a finite number of simple surfaces in τ. The proof already given actually covers this case. The only point which need be drawn to the reader's attention is that the integration which leads to equation (6.2) is still valid, because in the interval $z_P \leqslant z \leqslant z_Q$ the integrand $\partial F_3/\partial z$ will have at most a finite number of finite discontinuities; the Riemann integral of such a function certainly exists.[1]

[1] See, for example, E. G. Phillips: *A course of Analysis* (Cambridge), p. 174.

Although $\mathrm{div}\,\mathbf{F}$ will not be defined on a surface where the partial derivatives of $\mathbf{F}$ are not continuous, it is nevertheless convenient to leave the statement (6.1) unchanged. More accurately, we should replace $\mathrm{div}\,\mathbf{F}$ by $\partial F_1/\partial x + \partial F_2/\partial y + \partial F_3/\partial z$ on a surface of discontinuity.

Examples of regions covered by the proof. Two particular regions to which the divergence theorem applies are worthy of mention.

First, the theorem applies to the region bounded by a torus (or 'anchor ring'). For, take axes $Oxyz$ such that O is at the centre of the torus and Oz is perpendicular to its central plane (Fig. 69). The quadrant of the torus which lies in the region $x \geqslant 0$, $y \geqslant 0$ is bounded by a semi-convex surface; and the remaining portion can be cut into three similar quadrants, each piece being bounded by a semi-convex surface. By part (d) of the proof, it follows that the divergence theorem is valid for the torus.

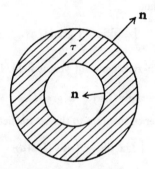

FIG. 70. The region τ between two spheres

Second, the divergence theorem applies to the region τ bounded by two concentric spheres. However, the integral

$$\int_S \mathbf{F}.\mathbf{dS}$$

is taken over both the inner and the outer bounding surfaces, and on each of these surfaces the unit normal is directed away from the enclosed region τ (Fig. 70). These facts can be established by taking axes $Oxyz$ with origin at the common centre of the spheres, and applying the divergence theorem to the portion of τ lying in each octant. We leave the verification to the reader.

EXAMPLE 1. The vector fields $\mathbf{F}, \mathbf{G}$ have continuous first derivatives throughout a closed region τ bounded by a simple closed surface S. At each point on S, $\mathbf{F} \times \mathbf{G}$ is tangential to the surface. Prove that

$$\int_\tau \mathbf{F}.\mathrm{curl}\,\mathbf{G}\,d\tau = \int \mathbf{G}.\mathrm{curl}\,\mathbf{F}\,d\tau.$$

Solution. Consider the identity (4.52), namely

$$\operatorname{div}(\mathbf{F} \times \mathbf{G}) = \mathbf{G} . \operatorname{curl} \mathbf{F} - \mathbf{F} . \operatorname{curl} \mathbf{G}.$$

As the conditions for the divergence theorem are fulfilled,

$$\int_S (\mathbf{F} \times \mathbf{G}) . d\mathbf{S} = \int_\tau \operatorname{div}(\mathbf{F} \times \mathbf{G}) \, d\tau$$

$$= \int_\tau (\mathbf{G} . \operatorname{curl} \mathbf{F} - \mathbf{F} . \operatorname{curl} \mathbf{G}) \, d\tau.$$

But $d\mathbf{S} = \mathbf{n} \, dS$, and as $\mathbf{F} \times \mathbf{G}$ is everywhere tangential to S (and so normal to $\mathbf{n}$) we have $(\mathbf{F} \times \mathbf{G}) . \mathbf{n} = 0$. Thus the surface integral on the left-hand side vanishes, and the required result follows.

EXAMPLE 2. (*Important corollary to the divergence theorem.*) If the scalar field Ω together with its gradient are defined throughout a closed region τ bounded by a simple closed surface S, then

$$\int_S \Omega \, d\mathbf{S} = \int_\tau \operatorname{grad} \Omega \, d\tau. \qquad (6.10)$$

Proof. Let $\mathbf{a}$ be any non-zero constant vector field. From the identity (4.49), as $\mathbf{a}$ is a constant vector,

$$\operatorname{div}(\Omega \mathbf{a}) = \mathbf{a} . \operatorname{grad} \Omega.$$

Thus applying the divergence theorem to the vector field $\Omega \mathbf{a}$,

$$\int_S \Omega \mathbf{a} . d\mathbf{S} = \int_\tau \operatorname{div}(\Omega \mathbf{a}) \, d\tau$$

$$= \int_\tau \mathbf{a} . \operatorname{grad} \Omega \, d\tau.$$

Since $\mathbf{a}$ is constant, we may rewrite this as

$$\mathbf{a} . \left(\int_S \Omega \, d\mathbf{S} - \int_\tau \operatorname{grad} \Omega \, d\tau \right) = 0.$$

Putting

$$\int_S \Omega \, d\mathbf{S} - \int_\tau \operatorname{grad} \Omega \, d\tau = \mathbf{F},$$

it follows that the component of $\mathbf{F}$ parallel to $\mathbf{a}$ is zero. But the direction of $\mathbf{a}$ can be chosen arbitrarily, and so $\mathbf{F} = \mathbf{0}$, which gives the required result.

EXERCISES

1. Using the divergence theorem, evaluate

$$\int_S \mathbf{F} . d\mathbf{S},$$

where $\mathbf{F} = (x^3, y^3, z^3)$ and S is the sphere $x^2 + y^2 + z^2 = a^2$. Verify the result by evaluating the surface integral directly.

2. If the scalar field Ω has continuous second-order derivatives in the closed region τ bounded by a simple closed surface S, prove that

$$\int_S \Omega \nabla \Omega . \, d\mathbf{S} = \int_\tau \{\Omega \nabla^2 \Omega + (\nabla \Omega)^2\} \, d\tau.$$

If $\Omega = x + y + z$, deduce that

$$\int_S \Omega \nabla \Omega . \, d\mathbf{S} = 3V,$$

where V is the volume of the region τ.

3. Show that

$$\int_S \mathbf{F} \times d\mathbf{S} = - \int_\tau \operatorname{curl} \mathbf{F} \, d\tau,$$

stating the conditions to be satisfied by $\mathbf{F}$ and the relationship of S and τ. [*Hint.* Apply the divergence theorem to the vector field $\mathbf{F} \times \mathbf{a}$, where $\mathbf{a}$ is an arbitrary constant vector field.]

4. If the region τ is bounded by a simple closed surface S, and $\mathbf{F}$, $\mathbf{G}$ are continuously differentiable in τ, prove that

$$\int_S \mathbf{F}(\mathbf{G}.\, d\mathbf{S}) = \int_\tau (\mathbf{G}.\nabla)\mathbf{F} \, d\tau + \int_\tau \mathbf{F} \operatorname{div} \mathbf{G} \, d\tau.$$

[*Hint.* Consider a single component of this vector equation.]

5. A region τ is bounded by simple closed surfaces S_1 and S_2. If the origin lies outside τ, prove that

$$\int_{S_1} \frac{\mathbf{r}.\, d\mathbf{S}_1}{r^3} + \int_{S_2} \frac{\mathbf{r}.\, d\mathbf{S}_2}{r^3} = 0.$$

6. A given continuously differentiable scalar function λ is positive in a region τ bounded by a simple closed surface S. Continuously differentiable vector fields $\mathbf{u}, \mathbf{v}, \mathbf{A}$, defined in τ, are such that:
   (i) $\mathbf{v} = \mathbf{u} + \operatorname{curl} \mathbf{A}$;
   (ii) $\operatorname{curl} \lambda \mathbf{u} = \mathbf{0}$;
   (iii) the tangential components of $\mathbf{A}$ vanish on S.
Prove that

$$\int_\tau \lambda v^2 \, d\tau \geqslant \int_\tau \lambda u^2 \, d\tau.$$

[*Hint.* Consider the integral over S of $2\lambda \mathbf{u} \times \mathbf{A}$.]

7. The scalar field Ω satisfies Laplace's equation in a region τ bounded by a simple closed surface S. Show that

$$\int_S \mathbf{n}.\operatorname{grad} \Omega \, dS = 0.$$

If S_r is a sphere of radius r and fixed centre O lying in τ, show that

$$\frac{d}{dr}\left[\frac{1}{r^2}\int_{S_r}\Omega\,dS\right] = \frac{1}{r^2}\int_{S_r}\frac{\partial\Omega}{\partial r}\,dS.$$

Taking spherical polar coordinates r, θ, ϕ with origin at O, and writing $\Omega(r,\theta,\phi)=\Omega_r$, deduce that

$$\int_0^{2\pi}\int_0^\pi (\Omega_r-\Omega_0)\sin\theta\,d\theta\,d\phi = 0.$$

Hence prove that the maximum and minimum values of Ω occur on the boundary S.

6.3 Green's theorems

Let scalar fields Φ and Ψ, together with $\nabla^2\Phi$ and $\nabla^2\Psi$, be defined throughout a closed region τ, bounded by a simple closed surface S, and suppose that any discontinuities in the second derivatives of Φ and Ψ are finite and are confined to a finite number of simple surfaces in τ. *Green's first theorem* is that

$$\int_S \Phi\frac{\partial\Psi}{\partial n}\,dS = \int_\tau (\Phi\nabla^2\Psi+\operatorname{grad}\Phi.\operatorname{grad}\Psi)\,d\tau. \tag{6.11}$$

Here, $\partial/\partial n$ denotes the directional derivative (see p. 81) along the outward normal to S. Interchanging Φ and Ψ in (6.11) we obtain

$$\int_S \Psi\frac{\partial\Phi}{\partial n}\,dS = \int_\tau (\Psi\nabla^2\Phi+\operatorname{grad}\Psi.\operatorname{grad}\Phi)\,d\tau;$$

whence, by subtraction from (6.11),

$$\int_S \left(\Phi\frac{\partial\Psi}{\partial n}-\Psi\frac{\partial\Phi}{\partial n}\right)dS = \int_\tau (\Phi\nabla^2\Psi-\Psi\nabla^2\Phi)\,d\tau, \tag{6.12}$$

which is *Green's second theorem*.

The first theorem is proved as follows:

Proof. Take $\mathbf{H}=\Phi\operatorname{grad}\Psi$. Then using identity (4.49),

$$\operatorname{div}\mathbf{H} = \Phi\operatorname{div}(\operatorname{grad}\Psi)+\operatorname{grad}\Psi.\operatorname{grad}\Phi$$
$$= \Phi\nabla^2\Psi+\operatorname{grad}\Phi.\operatorname{grad}\Psi.$$

Applying the divergence theorem to the vector field $\mathbf{H}$, we thus have

$$\int_S \Phi\operatorname{grad}\Psi.\mathbf{dS} = \int_\tau (\Phi\nabla^2\Psi+\operatorname{grad}\Phi.\operatorname{grad}\Psi)\,d\tau.$$

But $\Phi\operatorname{grad}\Psi.\mathbf{dS}=\Phi\mathbf{n}.\operatorname{grad}\Psi\,dS=\Phi(\partial\Psi/\partial n)\,dS$. Hence (6.11) follows.

LAPLACE'S EQUATION: A UNIQUENESS THEOREM. Let τ be a closed region, bounded by a simple surface S, and suppose a scalar field Ω is such that:

(i) Ω satisfies Laplace's equation in τ;   (6.13)

(ii) the values of Ω on S are given.   (6.14)

Then Ω is unique.

Proof. Suppose there are two scalar fields Ω, say Ω_1 and Ω_2, satisfying conditions (i) and (ii). Then

$$\nabla^2 \Omega_1 = 0, \quad \nabla^2 \Omega_2 = 0 \quad \text{throughout } \tau;$$

and

$$\Omega_1 = \Omega_2 \quad \text{at all points on } S.$$

Consider

$$\Phi = \Omega_1 - \Omega_2. \tag{6.15}$$

Then, since ∇^2 is a *linear* operator,

$$\nabla^2 \Phi = \nabla^2 \Omega_1 - \nabla^2 \Omega_2$$
$$= 0 \quad \text{throughout } \tau;$$

and

$$\Phi = 0 \text{ on } S.$$

Applying Green's theorem (6.11) with $\Psi = \Phi$, we thus have

$$\int_\tau (\text{grad}\,\Phi)^2 \, d\tau = 0. \tag{6.16}$$

Now $(\text{grad}\,\Phi)^2 \geqslant 0$ at all points. Suppose that $(\text{grad}\,\Phi)^2 > 0$ at some point P in the interior of τ, then as $\text{grad}\,\Phi$ is continuous throughout τ (the existence of $\text{grad}\,\Phi$ is sufficient to ensure this) there must be a region, τ_1 say, containing P and in which $(\text{grad}\,\Phi)^2 > 0$. We then have

$$\int_{\tau_1} (\text{grad}\,\Phi)^2 \, d\tau_1 > 0.$$

As the integral of $(\text{grad}\,\Phi)^2$ over the remainder of τ cannot be negative, it follows that

$$\int_\tau (\text{grad}\,\Phi)^2 \, d\tau > 0,$$

which contradicts (6.16). We conclude that at all points in τ, $\text{grad}\,\Phi = \mathbf{0}$. In components this becomes

$$\frac{\partial \Phi}{\partial x} = 0, \quad \frac{\partial \Phi}{\partial y} = 0, \quad \frac{\partial \Phi}{\partial z} = 0,$$

showing that Φ is independent of x, y and z. Thus, in τ, $\Phi = \text{constant}$. Since $\Phi = 0$ on the boundary of τ, the constant is zero, and so $\Phi \equiv 0$ in τ. Returning to (6.15), we see that this implies that $\Omega_1 \equiv \Omega_2$, which establishes the required result that there cannot be two different scalar fields each satisfying conditions (6.13) and (6.14).

Extension to infinite regions. Suppose that τ is the infinite region *outside* a simple closed surface S. The scalar field Ω is again unique if, in addition to conditions (6.13) and (6.14),

$$\Omega = O\left(\frac{1}{r}\right) \text{ at large distances } r \text{ (see footnote).} \tag{6.17}$$

Proof. We proceed as before, and consider scalar fields Ω_1 and Ω_2, each satisfying conditions (6.13), (6.14), and (6.17). Since $\Omega_1 = O(1/r)$ and $\Omega_2 = O(1/r)$ as $r \to \infty$, we also have

$$\Phi = \Omega_1 - \Omega_2 = O\left(\frac{1}{r}\right) \quad \text{as } r \to \infty. \tag{6.18}$$

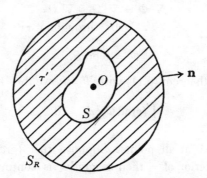

Fig. 71. The sphere S_R of radius R and centre the origin completely surrounds S. As $R \to \infty$ the sphere S_R becomes infinitely large.

Applying Green's theorem (6.11), with $\Phi = \Psi$, to the region τ' bounded internally by S and externally by a sphere S_R of radius R and centre at the origin O (Fig. 71),

$$\int_{\tau'} \{\Phi\nabla^2\Phi + (\operatorname{grad}\Phi)^2\}\, d\tau' = \int_S \Phi\frac{\partial\Phi}{\partial n}\, dS + \int_{S_R} \Phi\frac{\partial\Phi}{\partial n}\, dS_R.$$

As before, $\nabla^2\Phi = 0$ in τ' and $\Phi = 0$ on S. Also, on S_R, $\partial/\partial n \equiv \partial/\partial r$, because the outward unit normal to S_R is in the same direction as the unit vector $\hat{\mathbf{r}}$ from the origin. Thus

$$\int_{\tau'} (\operatorname{grad}\Phi)^2\, d\tau' = \int_{S_R} \left[\Phi\frac{\partial\Phi}{\partial r}\right]_{r=R} dS_R.$$

Expression (6.17) means that Ω tends to zero at least as rapidly as $1/r$ when $r \to \infty$. More precisely, there exist a value R of r and a positive constant K, independent of r, such that $|\Omega| \leqslant K/r$ whenever $r \geqslant R$.

If R is sufficiently large, then by (6.18) there exists a positive constant K such that for all $r \geqslant R$,

$$\left| \Phi \frac{\partial \Phi}{\partial r} \right| \leqslant \frac{K}{r} \times \frac{K}{r^2} = \frac{K^2}{r^3}.$$

Thus

$$\left| \int_{\tau'} (\operatorname{grad} \Phi)^2 \, d\tau' \right| \leqslant \int_{S_R} (K^2/R^3) \, dS_R$$

$$= 4\pi K^2/R,$$

since the area of S_R is $4\pi R^2$. When $R \to \infty$, τ' becomes the given infinite region τ, and so in the limit

$$\int_{\tau} (\operatorname{grad} \Phi)^2 \, d\tau = 0,$$

which is the same as equation (6.16). From this point the proof is completed as before.

EXERCISES

8. Assuming that τ is a closed region bounded by a simple closed surface S, obtain from Green's second theorem the following results:

(i) For any scalar field Φ having continuous second derivatives throughout τ,

$$\int_S \frac{\partial \Phi}{\partial n} \, dS = \int_{\tau} \nabla^2 \Phi \, d\tau.$$

(ii) For any scalar fields Φ and Ψ which satisfy Laplace's equation in τ,

$$\int_S \Phi \frac{\partial \Psi}{\partial n} \, dS = \int_S \Psi \frac{\partial \Phi}{\partial n} \, dS.$$

9. Scalar fields Φ, Φ' satisfy Laplace's equation in a region τ bounded by n simple closed surfaces $S_1, S_2, \ldots, S_n$. On the surface S_i $(i = 1, 2, \ldots, n)$,

$$\int_{S_i} \frac{\partial \Phi}{\partial n} \, dS_i = \alpha_i, \quad \int_{S_i} \frac{\partial \Phi'}{\partial n} \, dS_i = \alpha_i',$$

and

$$\Phi = C_i, \quad \Phi' = C_i',$$

where C_i, C_i' are constants. Using Green's second theorem, show that

$$\sum_{i=1}^{n} C_i \alpha_i' = \sum_{i=1}^{n} C_i' \alpha_{:}.$$

10. The scalar field Ω satisfies Laplace's equation in the closed region τ bounded by a simple closed surface S. If $\Omega = C$ (a constant) on S, prove by means of Green's first theorem (with $\Phi = \Psi = \Omega$) that $\Omega = C$ throughout τ. [*Note*. This result can also be deduced at once from the uniqueness theorem, proved in the text of this section.]

11. A closed region τ is bounded by a simple closed surface S. A continuously differentiable scalar field Ω is such that:
  (i) $\nabla^2 \Omega = 0$ in τ;
  (ii) $\partial\Omega/\partial n = f$ on S, where f is a given function and $\partial/\partial n$ denotes the derivative along the normal to S;
  (iii) Ω takes a given value at one point in τ.
Prove that Ω is unique.

If τ is an infinite region bounded internally by a simple closed surface S, and if condition (iii) is replaced by:
  (iii)' $\Omega = O(1/r)$ at large distances from S,
prove that Ω is again unique. [The proofs of these results follow closely the lines of the proof of the uniqueness theorem in the text of this section.]

6.4 Stokes's theorem

Consider a simple closed curve $\mathscr{C}$ spanned by a surface S, as in Fig. 72. Then, briefly (a complete statement is given later), Stokes's theorem states that, for a vector field $\mathbf{F}$,

$$\oint_{\mathscr{C}} \mathbf{F} \cdot d\mathbf{r} = \int_S \operatorname{curl} \mathbf{F} \cdot d\mathbf{S}. \tag{6.19}$$

For the theorem to hold, the orientation of the surface S and the curve $\mathscr{C}$ must be correctly related, and $\mathbf{F}$ must satisfy certain analytical conditions. The theorem will be proved first for curves spanned by flat surfaces and later will be extended to twisted curves spanned by surfaces of arbitrary shape.

Orientation of a surface spanning a closed curve. Let Q, R be neighbouring points on the boundary $\mathscr{C}$ of a simple open surface S, such that $\mathscr{C}$ is described in the sense of the smaller arc from Q to R (Fig. 73). Let P be a point on S near to Q and R. The surface S and its boundary $\mathscr{C}$ are said to be *correspondingly oriented* if $\overrightarrow{PQ}$, $\overrightarrow{PR}$ and the unit normal $\mathbf{n}$ to S (in that order) form a right-handed triad.

An equivalent statement is that S and $\mathscr{C}$ are correspondingly oriented if an observer on the positive side of S travelling around $\mathscr{C}$ in its positive sense always has the surface S on his left.

Stokes's theorem in the plane. Let $F_1(x,y)$ and $F_2(x,y)$ be defined and have continuous first derivatives throughout a closed region $\mathscr{R}$ in the xy-plane.

Let the boundary of $\mathscr{R}$ be a simple closed curve described in the anti-clockwise sense (Fig. 74). Then

$$\oint_{\mathscr{C}} \left(F_1 \frac{dx}{ds} + F_2 \frac{dy}{ds} \right) ds = \iint_{\mathscr{R}} \left(\frac{\partial F_2}{\partial x} - \frac{\partial F_1}{\partial y} \right) dx\, dy, \qquad (6.20)$$

where s denotes arc length along $\mathscr{C}$. This is Stokes's theorem in the plane. The reader should verify that (6.19) reduces to (6.20) when the surface S lies entirely in the xy-plane.[1]

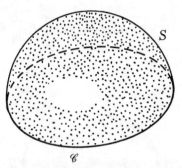

FIG. 72. A simple closed curve $\mathscr{C}$ spanned by a surface S

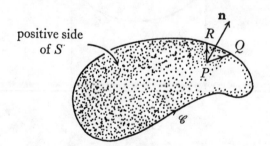

FIG. 73. The orientation of a surface S spanning a curve $\mathscr{C}$

Proof. We consider various cases in turn.

(a) *The case when $\mathscr{C}$ is such that any line parallel to the x- or y-axis has at most two points in common with $\mathscr{C}$.* Suppose that, in $\mathscr{R}$, x ranges from a to b, and let the ordinates through the points $(a, 0)$, $(b, 0)$ touch $\mathscr{C}$ at A, B,

---

[1] Stokes's theorem in the plane is often expressed more briefly as

$$\oint_{\mathscr{C}} (F_1\, dx + F_2\, dy) = \iint_{\mathscr{R}} \left(\frac{\partial F_2}{\partial x} - \frac{\partial F_1}{\partial y} \right) dx\, dy,$$

which implies that the curve $\mathscr{C}$ can be expressed in both the forms $y = f(x)$ and $x = g(y)$. However, as the curve is closed, f and g would not be single-valued functions. Therefore we shall retain the parameter s.

respectively. Let $y=y_1(x)$ and $y=y_2(x)$ be the equations of the lower and upper parts of $\mathscr{C}$ (between A and B) respectively (Fig. 74). Then

$$\iint\limits_{\mathscr{R}} -\frac{\partial F_1}{\partial y}\, dx\, dy = \int\limits_a^b \int\limits_{y_1(x)}^{y_2(x)} -\frac{\partial F_1}{\partial y}\, dy\, dx$$

$$= \int\limits_a^b \{F_1[x, y_1(x)] - F_1[x, y_2(x)]\}\, dx$$

$$= \int\limits_A^B F_1 \frac{dx}{ds}\, ds + \int\limits_B^A F_1 \frac{dx}{ds}\, ds, \qquad (6.21)$$

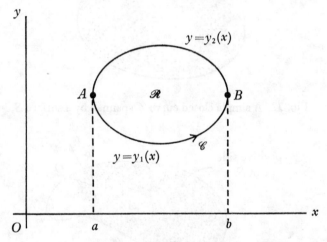

FIG. 74

where the first line integral is taken along the lower part of $\mathscr{C}$ and the second along the upper part. Thus

$$\iint\limits_{\mathscr{R}} -\frac{\partial F_1}{\partial y}\, dx\, dy = \oint\limits_{\mathscr{C}} F_1 \frac{dx}{ds}\, ds. \qquad (6.22)$$

Similarly, we can show that

$$\iint\limits_{\mathscr{R}} \frac{\partial F_2}{\partial x}\, dx\, dy = \oint\limits_{\mathscr{C}} F_2 \frac{dy}{ds}\, ds. \qquad (6.23)$$

Addition of the two results (6.22) and (6.23) thus gives (6.20), as required.

(b) *The case when part of $\mathscr{C}$ is parallel to an axis.* Suppose the curve $\mathscr{C}$ is as shown in Fig. 75, and let A, B, C be the points indicated on the diagram. We now obtain instead of (6.21)

$$\iint_{\mathcal{R}} -\frac{\partial F_1}{\partial y}\,dx\,dy = \int_A^B F_1 \frac{dx}{ds}\,ds + \int_B^C F_1 \frac{dx}{ds}\,ds$$

$$= \oint_{\mathscr{C}} F_1 \frac{dx}{ds}\,ds - \int_C^A F_1 \frac{dx}{ds}\,ds,$$

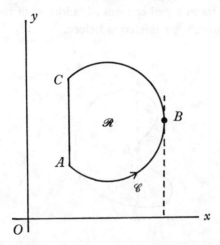

FIG. 75. The portion AC of $\mathscr{C}$ is parallel to the y-axis

where all the line integrals are taken along $\mathscr{C}$. But on CA, $dx/ds = 0$, and so the final integral vanishes. We thus have (6.22) once again, and the theorem follows as before.

(c) *The case when some lines parallel to the axes cut the boundary of $\mathcal{R}$ more than twice.* For simplicity we shall consider only the situation shown in Fig. 76, but the method can readily be extended to deal with still more complicated boundaries.

Subdivide $\mathcal{R}$ into regions $\mathcal{R}_1$, $\mathcal{R}_2$, $\mathcal{R}_3$ by drawing the line ABC, as in the diagram overleaf. Each of $\mathcal{R}_1$, $\mathcal{R}_2$, $\mathcal{R}_3$ is a region of the type considered in part (b). Denoting the boundary of $\mathcal{R}_1$ by $\mathscr{C}_1$, application of part (b) gives

$$\iint_{\mathcal{R}_1} -\frac{\partial F_1}{\partial y}\,dx\,dy = \oint_{\mathscr{C}_1} F_1 \frac{dx}{ds}\,ds$$

$$= \int_A^C F_1 \frac{dx}{ds}\,ds, \quad \text{taken along } \mathscr{C},$$

since $dx/ds = 0$ on the line CA. Similarly,

$$\iint_{\mathscr{R}_2} -\frac{\partial F_1}{\partial y}\, dx\, dy = \int_C^B F_1 \frac{dx}{ds}\, ds, \quad \text{taken along } \mathscr{C};$$

and

$$\iint_{\mathscr{R}_3} -\frac{\partial F_1}{\partial y}\, dx\, dy = \int_B^A F_1 \frac{dx}{ds}\, ds, \quad \text{taken along } \mathscr{C}.$$

Since $\mathscr{R}_1$, $\mathscr{R}_2$, $\mathscr{R}_3$ taken together form $\mathscr{R}$, addition of these three results gives (6.22). The proof is completed as before.

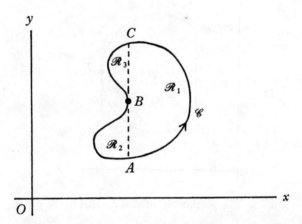

FIG. 76. Showing how to deal with more complicated boundaries

Lemma. If the vector fields $\mathbf{F} = \mathbf{F}(u, v)$ and $\mathbf{r} = \mathbf{r}(u, v)$ have continuous first partial derivatives, then

$$(\text{curl}\,\mathbf{F}).(\mathbf{r}_u \times \mathbf{r}_v) = \mathbf{r}_v.\mathbf{F}_u - \mathbf{r}_u.\mathbf{F}_v, \qquad (6.24)$$

where suffixes refer to partial derivatives with respect to u, v.

Proof. Using the notation of § 4.9, p. 94,

$$(\text{curl}\,\mathbf{F}).(\mathbf{r}_u \times \mathbf{r}_v) = \left(\mathbf{e}_i \times \frac{\partial \mathbf{F}}{\partial x_i}\right).\left(\frac{\partial \mathbf{r}}{\partial u} \times \frac{\partial \mathbf{r}}{\partial v}\right)$$

$$= \mathbf{e}_i.\left\{\frac{\partial \mathbf{F}}{\partial x_i} \times \left(\frac{\partial \mathbf{r}}{\partial u} \times \frac{\partial \mathbf{r}}{\partial v}\right)\right\}$$

$$= \mathbf{e}_i.\left\{\left(\frac{\partial \mathbf{r}}{\partial v}.\frac{\partial \mathbf{F}}{\partial x_i}\right)\frac{\partial \mathbf{r}}{\partial u} - \left(\frac{\partial \mathbf{r}}{\partial u}.\frac{\partial \mathbf{F}}{\partial x_i}\right)\frac{\partial \mathbf{r}}{\partial v}\right\}.$$

Now

$$\mathbf{r} = x_1\mathbf{e}_1 + x_2\mathbf{e}_2 + x_3\mathbf{e}_3 = x_j\mathbf{e}_j,$$

and so

$$\begin{aligned}
\mathbf{e}_i.(\partial\mathbf{r}/\partial u) &= \mathbf{e}_i.\mathbf{e}_j\,\partial x_j/\partial u \\
&= \delta_{ij}\,\partial x_j/\partial u \\
&= \partial x_i/\partial u.
\end{aligned}$$

Similarly,

$$\mathbf{e}_i.(\partial\mathbf{r}/\partial v) = \partial x_i/\partial v.$$

Thus

$$(\operatorname{curl}\mathbf{F}).(\mathbf{r}_u \times \mathbf{r}_v) = \left(\mathbf{r}_v.\frac{\partial\mathbf{F}}{\partial x_i}\right)\frac{\partial x_i}{\partial u} - \left(\mathbf{r}_u.\frac{\partial\mathbf{F}}{\partial x_i}\right)\frac{\partial x_i}{\partial v}.$$

Expanding the first term on the right of the above expression, and using the chain rule (4.3),

$$\begin{aligned}
\left(\mathbf{r}_v.\frac{\partial\mathbf{F}}{\partial x_i}\right)\frac{\partial x_i}{\partial u} &= \mathbf{r}_v.\left(\frac{\partial\mathbf{F}}{\partial x_1}\frac{\partial x_1}{\partial u} + \frac{\partial\mathbf{F}}{\partial x_2}\frac{\partial x_2}{\partial u} + \frac{\partial\mathbf{F}}{\partial x_3}\frac{\partial x_3}{\partial u}\right) \\
&= \mathbf{r}_v.\mathbf{F}_u.
\end{aligned}$$

Using this and a similar expression with u and v interchanged, it follows that

$$(\operatorname{curl}\mathbf{F}).(\mathbf{r}_u \times \mathbf{r}_v) = \mathbf{r}_v.\mathbf{F}_u - \mathbf{r}_u.\mathbf{F}_v,$$

as required.

With the help of the above lemma and the knowledge of Stokes's theorem in the plane, we can now proceed to the general case, as follows.

Stokes's theorem. Let the vector field $\mathbf{F}$, together with $\operatorname{curl}\mathbf{F}$, be defined everywhere on a simple open surface S with *correspondingly oriented* boundary $\mathscr{C}$. Then

$$\oint_{\mathscr{C}} \mathbf{F}.d\mathbf{r} = \int_{S} \operatorname{curl}\mathbf{F}.d\mathbf{S}. \tag{6.25}$$

Proof. Let the parametric equation of S be

$$\mathbf{r} = \mathbf{r}(u, v), \tag{6.26}$$

where u, v range over a closed region $\mathscr{R}$ in the uv-plane. Apart from the usual analytical restrictions on $\mathbf{r}(u,v)$ and its first partial derivatives, we shall find it necessary in the course of the proof to assume that the second partial derivatives of $\mathbf{r}$ with respect to u, v exist and are continuous. This is a condition on the nature of the surface S slightly stronger than usual, but in practical applications this will cause no difficulty. We denote by $\mathscr{C}'$ the boundary of $\mathscr{R}$ (Fig. 77).

The boundary $\mathscr{C}'$ of $\mathscr{R}$ is mapped into the boundary $\mathscr{C}$ of S by (6.26). For, as the mapping is single-valued and continuous, if $\mathscr{C}'$ is shrunk continuously to a point P' in $\mathscr{R}$ the corresponding curve $\mathscr{C}$ will sweep out in a continuous manner the whole of S, and will finally reduce to a point P on S.

If $\mathscr{C}'$ is traversed in the anti-clockwise sense (Fig. 77), the surface S and the boundary $\mathscr{C}$ will be correspondingly oriented if we take the unit normal to S as

$$\mathbf{n} = \mathbf{r}_u \times \mathbf{r}_v / |\mathbf{r}_u \times \mathbf{r}_v|.$$

To see this, take any point P' in $\mathscr{R}$ close to $\mathscr{C}'$, such that the lines through P' in the directions u increasing and v increasing meet $\mathscr{C}'$ at neighbouring points Q' and R' respectively (Fig. 77). Let P, Q, R be the corresponding points on S. When $\mathscr{C}'$ is described in the anticlockwise sense from Q' to R', $\mathscr{C}$ will be described in the sense of the smaller arc from Q to R. Also, P and Q lie on the same u-coordinate curve, with u increasing towards Q;

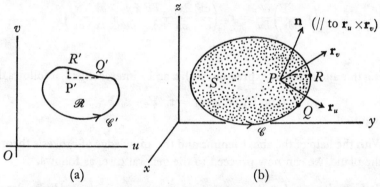

(a)   (b)

FIG. 77. (a) The mapping onto the uv-plane of the surface S shown in (b). If the surface normal $\mathbf{n}$ to S is chosen parallel to $\mathbf{r}_u \times \mathbf{r}_v$, $\mathscr{C}'$ will be traversed in the conventional anticlockwise sense

and P and R lie on the same v-coordinate curve, with v increasing towards R. As $\mathbf{r}_u$, $\mathbf{r}_v$ are tangential to the u-, v-coordinate curves and in the directions u, v increasing, respectively, and as $\mathbf{r}_u$, $\mathbf{r}_v$ and $\mathbf{n}$ (in that order) form a right-handed triad, it follows that $\overrightarrow{PQ}$, $\overrightarrow{PR}$, and $\mathbf{n}$ also form a right-handed triad. This is the required condition for $\mathscr{C}$ and S to be correspondingly oriented.

Consider the right-hand side of (6.25). Substituting $\mathbf{dS} = (\mathbf{r}_u \times \mathbf{r}_v) \, du \, dv$, using the lemma (6.24), and assuming that the second derivatives of $\mathbf{r}$ are continuous (so that $\mathbf{r}_{uv} = \mathbf{r}_{vu}$), we have

$$\int_S \operatorname{curl} \mathbf{F} \cdot \mathbf{dS} = \iint_{\mathscr{R}} (\mathbf{r}_v \cdot \mathbf{F}_u - \mathbf{r}_u \cdot \mathbf{F}_v) \, du \, dv$$

$$= \iint_{\mathscr{R}} \left\{ \frac{\partial}{\partial u} (\mathbf{F} \cdot \mathbf{r}_v) - \frac{\partial}{\partial v} (\mathbf{F} \cdot \mathbf{r}_u) \right\} du \, dv.$$

Applying Stokes's theorem in the plane, as given by (6.20), it follows that

$$\int_S \operatorname{curl}\mathbf{F}.\mathbf{dS} = \oint_{\mathscr{C}'} \left(\mathbf{F}.\mathbf{r}_u\frac{du}{ds'}+\mathbf{F}.\mathbf{r}_v\frac{dv}{ds'}\right)ds'$$

$$= \oint_{\mathscr{C}} \mathbf{F}.\left(\frac{\partial\mathbf{r}}{\partial u}\frac{du}{ds}+\frac{\partial\mathbf{r}}{\partial v}\frac{dv}{ds}\right)ds,$$

where s' and s denote arc lengths along $\mathscr{C}'$ and $\mathscr{C}$ respectively. But the chain rule (4.3) gives

$$\frac{\partial\mathbf{r}}{\partial u}\frac{du}{ds}+\frac{\partial\mathbf{r}}{\partial v}\frac{dv}{ds} = \frac{d\mathbf{r}}{ds}.$$

Hence, as required,

$$\int_S \operatorname{curl}\mathbf{F}.\mathbf{dS} = \oint_{\mathscr{C}} \mathbf{F}.\mathbf{dr}.$$

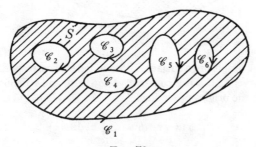

FIG. 78

Corollary: Surfaces with several boundaries. If the simple open surface S is bounded by n correspondingly oriented simple closed curves, $\mathscr{C}_1, \mathscr{C}_2, \ldots,$ $\mathscr{C}_n$ (Fig. 78 illustrates the case $n=6$) we have instead of (6.25):

$$\oint_{\mathscr{C}_1} \mathbf{F}.\mathbf{dr}+\oint_{\mathscr{C}_2} \mathbf{F}.\mathbf{dr}+\cdots+\oint_{\mathscr{C}_n} \mathbf{F}.\mathbf{dr} = \int_S \operatorname{curl}\mathbf{F}.\mathbf{dS}. \qquad (6.27)$$

Proof. Consider the case when there are just two boundaries $\mathscr{C}_1, \mathscr{C}_2$ (Fig. 79). Let AB be a simple curve on S (called a *cross cut*) joining a point A on $\mathscr{C}_1$ to a point B on $\mathscr{C}_2$. If S is cut along AB, we may regard the surface as having a single boundary consisting of $\mathscr{C}_1$, AB, $\mathscr{C}_2$, and BA as shown. Applying Stokes's theorem in the form already proved,

$$\oint_{\mathscr{C}_1} \mathbf{F}.\mathbf{dr}+\int_{AB} \mathbf{F}.\mathbf{dr}+\oint_{\mathscr{C}_2} \mathbf{F}.\mathbf{dr}+\int_{BA} \mathbf{F}.\mathbf{dr} = \int_S \operatorname{curl}\mathbf{F}.\mathbf{dS}.$$

But

$$\int_{AB} \mathbf{F}.\mathbf{dr} = -\int_{BA} \mathbf{F}.\mathbf{dr}.$$

Thus

$$\oint_{\mathscr{C}_1} \mathbf{F}.d\mathbf{r} + \oint_{\mathscr{C}_2} \mathbf{F}.d\mathbf{r} = \int_{S} \text{curl}\,\mathbf{F}.d\mathbf{S},$$

which proves (6.27) for the case $n = 2$.

The general case can be treated by introducing $n - 1$ suitable cross cuts.

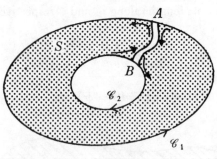

FIG. 79

EXAMPLE 3. A simple open surface S is bounded by a correspondingly oriented simple closed curve $\mathscr{C}$. If $\mathbf{a}$ is a constant vector field, and $\mathbf{r}$ denotes the position vector relative to the origin, prove that

$$\oint_{\mathscr{C}} (\mathbf{a} \times \mathbf{r}).d\mathbf{r} = 2 \int_{S} \mathbf{a}.d\mathbf{S}.$$

Deduce that

$$\int_{S} d\mathbf{S} = \tfrac{1}{2} \oint_{\mathscr{C}} \mathbf{r} \times d\mathbf{r}.$$

Solution. By Stokes's theorem,

$$\oint_{\mathscr{C}} (\mathbf{a} \times \mathbf{r}).d\mathbf{r} = \int_{S} \text{curl}(\mathbf{a} \times \mathbf{r}).d\mathbf{S} = I \text{ (say)}.$$

Using identity (4.53) and the fact that $\mathbf{a}$ is constant,

$$I = \int_{S} \{\mathbf{a}\,\text{div}\,\mathbf{r} - (\mathbf{a}.\nabla)\mathbf{r}\}.d\mathbf{S}.$$

If $\mathbf{a} = (a_1, a_2, a_3)$ relative to the axes $Oxyz$, we have

$$(\mathbf{a}.\nabla)\mathbf{r} = \left(a_1\frac{\partial}{\partial x} + a_2\frac{\partial}{\partial y} + a_3\frac{\partial}{\partial z}\right)(x, y, z)$$

$$= (a_1, a_2, a_3)$$

$$= \mathbf{a}.$$

Also, $\operatorname{div}\mathbf{r} = \operatorname{div}(x, y, z) = 3$. Thus

$$\oint_{\mathscr{C}} (\mathbf{a} \times \mathbf{r}).d\mathbf{r} = 2 \int_{S} \mathbf{a}.d\mathbf{S}.$$

Since $\mathbf{a}$ takes the same value at all points on $\mathscr{C}$ and S, the above result may be expressed as

$$\mathbf{a}.\left(\oint_{\mathscr{C}} \mathbf{r} \times d\mathbf{r} - 2 \int_{S} d\mathbf{S} \right) = 0.$$

By an argument similar to that used in Example 2 on p. 166, it follows that the bracketed expression vanishes, which gives the result required.

EXERCISES

12. Verify Stokes's theorem for the vector field $\mathbf{F} = (x^2 y, z, 0)$ and the hemisphere

$$x^2 + y^2 + z^2 = a^2, \quad z \geqslant 0.$$

13. If S is a simple open surface, bounded by a correspondingly oriented curve $\mathscr{C}$, and Φ, Ψ are continuously differentiable scalar fields, prove from Stokes's theorem that

$$\oint_{\mathscr{C}} \Phi \operatorname{grad} \Psi.d\mathbf{r} = \int_{S} \operatorname{grad} \Phi \times \operatorname{grad} \Psi.d\mathbf{S}.$$

14. A simple open surface S is bounded by a correspondingly oriented curve $\mathscr{C}$. If Ω is a continuously differentiable scalar field, prove the following corollary of Stokes's theorem:

$$\oint_{\mathscr{C}} \Omega \, d\mathbf{r} = - \int_{S} \operatorname{grad} \Omega \times d\mathbf{S}.$$

[*Hint.* Consider $\Omega\mathbf{a}$, where $\mathbf{a}$ is a constant vector field, and proceed as in Example 2, p. 166.]

15. Prove the following corollary of Stokes's theorem, and state the conditions to be satisfied by $\mathbf{F}$, S and $\mathscr{C}$:

$$\oint_{\mathscr{C}} \mathbf{F} \times d\mathbf{r} = - \int_{S} (\mathbf{n} \times \nabla) \times \mathbf{F} \, dS.$$

[*Hint.* Put $\mathbf{F} = F_1\mathbf{e}_1 + F_2\mathbf{e}_2 + F_3\mathbf{e}_3$, where $\mathbf{e}_1$, $\mathbf{e}_2$, $\mathbf{e}_3$ are the basic unit vectors of a rectangular cartesian coordinate system, and use the result of Exercise 14.]

16. Two non-intersecting simple closed curves $\mathscr{C}_1$, $\mathscr{C}_2$ lie in the xy-plane, with $\mathscr{C}_1$ enclosing $\mathscr{C}_2$, and are described in the anticlockwise sense. Show that, for the vector field $\mathbf{F} = (zy^2, 2xyz, x)$,

$$\oint_{\mathscr{C}_1} \mathbf{F}.d\mathbf{r} = \oint_{\mathscr{C}_2} \mathbf{F}.d\mathbf{r}.$$

17. Consider the remarks made in § 6.2 about the extension of Gauss's theorem to functions whose partial derivatives are not continuous. Verify that Stokes's

theorem (6.25) extends in a similar way, provided that any discontinuities in the derivatives of **F** are finite and are confined to a finite number of simple curves on the surface S.

6.5   Limit definitions of div F and curl F

Div **F**. Let P be a point in a region τ bounded by a simple closed surface S, throughout which the vector field **F** and its divergence are defined. If V denotes the volume of τ, the mean value theorem for integrals[1] shows that there exists a point P' in τ such that

$$\frac{1}{V} \int_{\tau} \operatorname{div} \mathbf{F} \, d\tau = \operatorname{div} \mathbf{F}, \quad \text{evaluated at } P'.$$

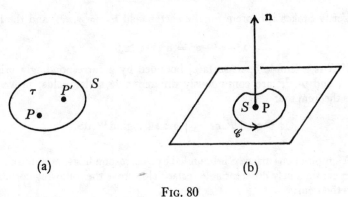

(a)   (b)

FIG. 80

Thus, by the divergence theorem,

$$\frac{1}{V} \int_{S} \mathbf{F} . \, d\mathbf{S} = \operatorname{div} \mathbf{F}, \quad \text{evaluated at } P.$$

If the surface S is now allowed to collapse towards P', the linear dimensions of τ become arbitrarily small, and P' must approach P (Fig. 80(a)). It follows that at P

$$\operatorname{div} \mathbf{F} = \lim_{V \to 0} \frac{1}{V} \int_{S} \mathbf{F} . \, d\mathbf{S}. \tag{6.28}$$

This relationship is very often used to define div **F**.

Curl **F**. Let **F** and curl **F** be defined throughout a region τ containing a given point P. Draw a plane surface S through P and contained in τ, with its normal in the direction of a given unit vector **n** and bounded by a correspondingly oriented simple closed curve $\mathscr{C}$ (Fig. 80 (b)). Let A be the area of

[1] See, for example, R. Courant, *Differential and Integral Calculus*, Vol. II (Blackie), p. 232.

the surface S. Then, using Stokes's theorem followed by the mean value theorem, there exists a point P' on S such that

$$\oint_{\mathscr{C}} \mathbf{F}.d\mathbf{r} = \int_{S} \mathbf{n}.\mathrm{curl}\,\mathbf{F}\,dS$$

$$= A\mathbf{n}.\mathrm{curl}\,\mathbf{F}, \quad \text{evaluated at } P'.$$

Allowing $\mathscr{C}$ to shrink towards P, it follows that at P

$$\mathbf{n}.\mathrm{curl}\,\mathbf{F} = \lim_{A \to 0} \frac{1}{A} \oint_{\mathscr{C}} \mathbf{F}.d\mathbf{r}. \tag{6.29}$$

This gives the component of $\mathrm{curl}\,\mathbf{F}$ at P in the direction of $\mathbf{n}$. As the direction of the unit vector $\mathbf{n}$ may be chosen arbitrarily, we can obtain the components of $\mathrm{curl}\,\mathbf{F}$ along three non-coplanar axes, and (6.29) thus effectively specifies the value of $\mathrm{curl}\,\mathbf{F}$ at P completely.

EXERCISES

18. Let S denote the surface of a sphere of variable radius, centred at the origin, and let V be the volume enclosed by the sphere. Verify that

$$(\mathrm{div}\,\mathbf{r})_O = \lim_{V \to 0} \frac{1}{V} \int_{S} \mathbf{r}.d\mathbf{S},$$

where the suffix O indicates the value at the origin.

19. Let $\mathscr{C}$ be the rectangular contour $x = \pm h$, $y = \pm h$, lying in the xy-plane and described in the anticlockwise sense. Verify that for the vector field $\mathbf{F} = (z^2, x, y^2)$,

$$\mathbf{k}.(\mathrm{curl}\,\mathbf{F})_O = \lim_{h \to 0} \frac{1}{4h^2} \oint_{\mathscr{C}} \mathbf{F}.d\mathbf{r},$$

where the suffix O indicates the value at the origin.

6.6 Geometrical and physical significance of divergence and curl

Let S be a closed surface bounding a region τ of volume V, and consider

$$I = \frac{1}{V} \int_{S} \mathbf{F}.d\mathbf{S}.$$

If, at each point on S, $\mathbf{F}$ is directed away from the enclosed region τ, then $\mathbf{F}.\mathbf{n} > 0$ at each point and $I > 0$. Similarly, if $\mathbf{F}$ points into τ at each point on S, then $I < 0$. In the first case $\mathbf{F}$ is diverging from τ and in the second case $\mathbf{F}$ is converging towards τ.

13

More generally, $\mathbf{F} \cdot \mathbf{n}$ may take some positive and some negative values on S. The sign of I indicates whether the average field on S is divergent or convergent. Allowing S to shrink towards a fixed point P and using (6.28), it follows that the sign of $\operatorname{div} \mathbf{F}$ shows whether the mean field in the neighbourhood of P is divergent or convergent; the magnitude of $\operatorname{div} \mathbf{F}$ gives an indication of the strength of the divergence or convergence. In the early days of vector analysis, the expression

$$-\frac{\partial F_1}{\partial x} - \frac{\partial F_2}{\partial y} - \frac{\partial F_3}{\partial z}$$

was often referred to as the *convergence* of the vector field $\mathbf{F} = (F_1, F_2, F_3)$.

The meaning of divergence can be clearly visualized in the theory of fluid mechanics. In a continuous motion of an *incompressible* fluid, there can be

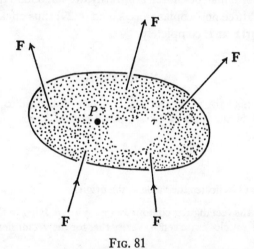

FIG. 81

no net flow of fluid into or out of any fixed closed surface (assuming that no cavities occur and that no fluid is being created or destroyed). Denoting the fluid velocity by $\mathbf{v}$,

$$\int_S \mathbf{v} \cdot d\mathbf{S}$$

gives the net flux across a surface S. We should thus expect $\operatorname{div} \mathbf{v}$ to vanish at all points, and this is indeed one of the field equations governing the flow of an incompressible fluid.

To understand the significance of $\operatorname{curl} \mathbf{F}$, take an *arbitrarily small* circular disc centred at a point P and with its axis in the direction of $\operatorname{curl} \mathbf{F}$ at P. Let $\mathscr{C}$ denote the boundary of the disc, and consider

$$I = \oint_{\mathscr{C}} \mathbf{F} \cdot d\mathbf{r} = \oint_{\mathscr{C}} \mathbf{F} \cdot \hat{\mathbf{T}} \, ds,$$

where $\hat{\mathbf{T}}$ is the unit tangent to $\mathscr{C}$. Since $\mathbf{n}$ (the unit normal to the disc) and curl $\mathbf{F}$ are in the same direction, $\mathbf{n} \cdot \text{curl}\,\mathbf{F} > 0$, and so $I > 0$. Averaged over the boundary of the disc, the value of $\mathbf{F} \cdot \hat{\mathbf{T}}$ is thus positive, and this indicates that the field in the neighbourhood of P has (on the average) a rotational component directed in the positive sense of $\mathscr{C}$ (Fig. 82). The field thus rotates about an axis parallel to curl $\mathbf{F}$. When curl $\mathbf{F} = \mathbf{0}$, there is no net rotation.

In fluid mechanics, the vanishing of curl $\mathbf{v}$ ($\mathbf{v}$ = velocity) implies that the fluid elements are not in a state of rotation. Such flows are termed *irrotational*.

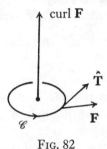

Fɪɢ. 82

EXERCISES

20. Illustrate the remarks made in this section about the significance of divergence by reference to the vector field $\mathbf{F} = (x, y, z)$, taking a spherical surface centred at the origin. Discuss in a similar way the divergence of the vector field $\mathbf{F} = (x^2, y^2, z^2)$ at the origin O.

21. Illustrate the remarks made in this section about curl $\mathbf{F}$ by reference to the vector fields

$$\text{(i)} \ \mathbf{F} = (x, y, z), \qquad \text{(ii)} \ \mathbf{F} = (y, 0, 0),$$

in each case choosing a rectangular circuit in the xy-plane.

22. A rigid body spins about a fixed axis through its centre O with angular velocity $\boldsymbol{\omega}$. If $\mathbf{v}$ is the velocity of any point in the body, prove that

$$\boldsymbol{\omega} = \tfrac{1}{2} \text{curl}\,\mathbf{v}.$$

[*Note.* Because of this connection with rotation, curl $\mathbf{v}$ is sometimes denoted by rot $\mathbf{v}$.]

CHAPTER 7

APPLICATIONS IN POTENTIAL THEORY

The applications of vector analysis lie mainly in electromagnetism, fluid mechanics, gravitation, and elasticity, and in each case scalar and vector potentials play an important part in the theory. In this chapter a brief account is given of potential functions and some closely related topics.

7.1 Connectivity

Let $\mathscr{C}$ and $\mathscr{C}'$ be simple closed curves drawn in a region $\mathscr{R}$. The curves are said to be *reconcilable* if it is possible to continuously distort one curve (without 'opening' it) so as to bring it into coincidence with the other, without the curve ever leaving $\mathscr{R}$.

A region is said to be *singly connected* (or simply connected) if all simple closed curves drawn in the region are reconcilable with each other.

If the simple closed curves which can be drawn in a region fall into two distinct classes, such that all the curves in one class are reconcilable with each other but not with curves in the other class, then the region is said to be *doubly connected* or to have *connectivity two*.

In general, a region $\mathscr{R}$ has *connectivity n* if there are just n classes of simple closed curves in $\mathscr{R}$, the curves in any class being reconcilable with each other but not with curves in other classes.

EXAMPLES. The interior of a sphere and the interior of an infinitely long circular cylinder are singly connected regions. However, the region outside a circular cylinder is doubly connected; and so is the region inside a torus (or anchor ring). In two dimensions, the region between two concentric circles is doubly connected. The reader should satisfy himself that these statements are in accordance with the definitions.

Reduction of a multiply connected region. A region which is multiply connected is often reduced to a singly connected region by removing all points on suitable surfaces.

For example, consider the region $\mathscr{R}$ between two coaxial cylinders C_1, C_2 (Fig. 83). Curves in $\mathscr{R}$ which surround the inner cylinder, C_1, are reconcilable with each other, and closed curves in $\mathscr{R}$ which do not surround C_1 are reconcilable with each other. As the curves in one class are not reconcilable with curves in the other, the region $\mathscr{R}$ is doubly connected. However, suppose that all the points on a plane AB joining a generator of C_1 to a generator of C_2 are removed. The remaining region, $\mathscr{R}'$ say, is then singly connected, because the only closed curves that lie entirely in $\mathscr{R}'$ do not surround C_1 (a closed curve surrounding C_1 passes outside $\mathscr{R}'$ where it crosses AB).

When a multiply connected region arises in practice, it is nearly always sufficient to confine attention to a corresponding singly connected region by introducing suitable 'cuts'. We shall therefore discuss only singly connected regions.

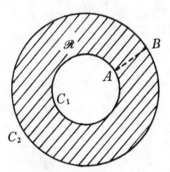

FIG. 83. The region $\mathscr{R}$ between coaxial cylinders can be made singly connected by removing from $\mathscr{R}$ all points of a plane joining the cylinders

EXERCISE

1. State the degree of connectivity (one, two, three, etc.) of the following:

 (i) the region outside a torus;
 (ii) the region outside two spheres, neither of which encloses or intersects the other;
 (iii) the two-dimensional region outside two coplanar circles, neither of which encloses or intersects the other.

7.2   The scalar potential

It was shown earlier (equation (4.47), p. 95) that if a vector field $\mathbf{F}$ is such that $\mathbf{F} = \operatorname{grad}\Omega$, then $\operatorname{curl}\mathbf{F} = 0$. We now prove the converse of this result, namely, that if a vector field $\mathbf{F}$ defined in a singly connected region $\mathscr{R}$ is such that

$$\operatorname{curl}\mathbf{F} \equiv 0 \text{ in } \mathscr{R}, \tag{7.1}$$

then there exists a scalar field Ω, defined in $\mathscr{R}$, such that

$$\mathbf{F} \equiv \operatorname{grad}\Omega. \tag{7.2}$$

The function Ω is called a *scalar potential* of the vector field $\mathbf{F}$.

Proof. Choose any point O in $\mathscr{R}$ as origin, and denote by P the point with coordinates (x, y, z). Consider the line integral

$$\Omega = \int_O^P \mathbf{F} . d\mathbf{r}, \tag{7.3}$$

taken along any simple curve in $\mathscr{R}$ joining O and P. We shall show that Ω is a scalar field with the required property that $\mathbf{F} = \operatorname{grad} \Omega$.

Let $\mathscr{C}$ be any simple closed curve in $\mathscr{R}$, passing through both O and P. Since $\operatorname{curl} \mathbf{F} \equiv \mathbf{0}$ in $\mathscr{R}$, it follows from Stokes' theorem (p. 177) that

$$\oint_{\mathscr{C}} \mathbf{F} . d\mathbf{r} = 0.$$

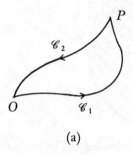 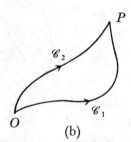

(a)   (b)

FIG. 84

If $\mathscr{C}$ consists of two portions $\mathscr{C}_1$, $\mathscr{C}_2$, as shown in Fig. 84(a), this becomes

$$\int_{\mathscr{C}_1 O}^P \mathbf{F} . d\mathbf{r} + \int_{\mathscr{C}_2 P}^O \mathbf{F} . d\mathbf{r} = 0.$$

However, if $\mathscr{C}_2$ is described in the sense from O to P, as in Fig. 84(b), we obtain

$$\int_{\mathscr{C}_1 O}^P \mathbf{F} . d\mathbf{r} - \int_{\mathscr{C}_2 O}^P \mathbf{F} . d\mathbf{r} = 0.$$

It follows that Ω, as defined by (7.3), is independent of the curve joining O to P, and is therefore a scalar function of x, y, z (the coordinates of P).

To show that $\mathbf{F} = \operatorname{grad} \Omega$, let $P(x_0, y_0, z_0)$ and $Q(x, y_0, z_0)$ be two points such that PQ lies wholly in $\mathscr{R}$. Then

$$\Omega(x, y_0, z_0) - \Omega(x_0, y_0, z_0) = \int_O^Q \mathbf{F} . d\mathbf{r} - \int_O^P \mathbf{F} . d\mathbf{r}$$

$$= -\int_Q^O \mathbf{F} . d\mathbf{r} - \int_O^P \mathbf{F} . d\mathbf{r}.$$

But

$$\int\limits_O^P \mathbf{F}.d\mathbf{r} + \int\limits_P^Q \mathbf{F}.d\mathbf{r} + \int\limits_Q^O \mathbf{F}.d\mathbf{r} = 0,$$

by Stokes's theorem. Thus

$$\Omega(x,y_0,z_0) - \Omega(x_0,y_0,z_0) = \int\limits_P^Q \mathbf{F}.d\mathbf{r}.$$

The curve along which the integral is taken can be the straight line PQ, which is parallel to Ox, and thus we have

$$\Omega(x,y_0,z_0) - \Omega(x_0,y_0,z_0) = \int\limits_{x_0}^x F_1\,dx,$$

where F_1 denotes the x-component of $\mathbf{F}$.[1] Differentiating with respect to x, as x_0, y_0, z_0 are constant,

$$F_1 = \partial\Omega/\partial x;$$

here we have used the fact that F_1 is continuous, which is implied by the existence of curl $\mathbf{F}$. We may similarly prove that the y- and z-components of $\mathbf{F}$ are respectively $F_2 = \partial\Omega/\partial y$ and $F_3 = \partial\Omega/\partial z$. Hence

$$\mathbf{F} = \operatorname{grad}\Omega,$$

as required.

Uniqueness. Suppose that Ω and Ω' are two scalar potentials such that

$$\mathbf{F} - \operatorname{grad}\Omega, \quad \mathbf{F} = \operatorname{grad}\Omega',$$

and write

$$U = \Omega - \Omega'.$$

By subtraction, grad $U = \mathbf{0}$, and hence, taking components,

$$\partial U/\partial x = 0, \quad \partial U/\partial y = 0, \quad \partial U/\partial z = 0.$$

These equations show that U is independent of x, y and z, giving $U = $ constant. Thus

$$\Omega' = \Omega + \text{constant}.$$

This shows that *the scalar potential is unique apart from an arbitrary additive constant.* Some degree of arbitrariness in Ω should be expected, because the integral (7.3) depends upon the origin O, whose position was chosen arbitrarily.

---

[1] The use of x as a variable in the integrand and in the limits is convenient and should cause no confusion here. We could, of course, replace x in the integrand by some other dummy variable, say t.

Irrotational vector fields. A vector field $\mathbf{F}$ such that $\operatorname{curl}\mathbf{F} \equiv \mathbf{0}$ in a region $\mathscr{R}$ is said to be *irrotational* in $\mathscr{R}$.

Conservative vector fields. A vector field $\mathbf{F}$ such that, for any two points P and Q,

$$\int_P^Q \mathbf{F}.d\mathbf{r}$$

is independent of the path of integration between P and Q is said to be *conservative.*

EXERCISES

2. Show that, in a singly connected region, an irrotational vector field is conservative; and, conversely, a conservative vector field is irrotational.

3. In Newtonian mechanics, when a force $\mathbf{F}$ is a function of position only, the *potential energy* of the force at any point P is defined as

$$-\int_O^P \mathbf{F}.d\mathbf{r},$$

where O is any origin. Deduce that the potential energy of the gravitational force on a particle of mass m is mgz, where g denotes acceleration due to gravity and z denotes height.

4. Find scalar potentials of the vector fields

$$\mathbf{F} = r^{-3}\mathbf{r} \quad \text{and} \quad \mathbf{F} = \mathbf{r}.$$

5. Verify that a scalar potential of the vector field

$$\mathbf{F} = 3(\boldsymbol{\mu}.\mathbf{r})r^{-5}\mathbf{r} - \boldsymbol{\mu}\,r^{-3}$$

is

$$\Omega = -(\boldsymbol{\mu}.\mathbf{r})/r^3,$$

where $\boldsymbol{\mu}$ is a constant vector.

7.3 The vector potential

On p. 95, equation (4.46), we showed that if, in a region $\mathscr{R}$, a vector field $\mathbf{F}$ having continuous second derivatives is such that $\mathbf{F} \equiv \operatorname{curl}\mathbf{A}$, then $\operatorname{div}\mathbf{F} \equiv 0$. We now prove the converse of this result, namely, that if

$$\operatorname{div}\mathbf{F} \equiv 0 \quad \text{in } \mathscr{R}, \tag{7.4}$$

then there exists a vector field $\mathbf{A}$ such that

$$\mathbf{F} \equiv \operatorname{curl}\mathbf{A} \quad \text{in } \mathscr{R}, \tag{7.5}$$

The function $\mathbf{A}$ is called a *vector potential* of the vector field $\mathbf{F}$.

Proof. Denote by

$$\xi(y, z)$$

the value of F_1, the x-component of $\mathbf{F}$, at the point $(0, y, z)$ of the plane $x = 0$. If F_2 and F_3 are the y- and z-components of $\mathbf{F}$, consider

$$\mathbf{A} = \left(\int_0^x F_3\, dx - \int_0^z \xi\, dz \right) \mathbf{j} - \left(\int_0^x F_2\, dx \right) \mathbf{k}, \tag{7.6}$$

where

$$\int_0^x F_2\, dx \quad \text{and} \quad \int_0^x F_3\, dx$$

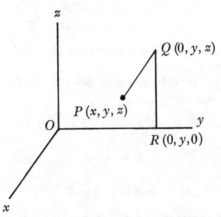

Fig. 85

are the integrals of $F_2(x, y, z)$ and $F_3(x, y, z)$ with respect to x (y and z being fixed) from the point $Q(0, y, z)$ to the point $P(x, y, z)$; and

$$\int_0^z \xi\, dz$$

is the integral of $\xi(y, z) = F_1(0, y, z)$ with respect to z from the point $R(0, y, 0)$ to the point $Q(0, y, z)$ (Fig. 85).[1] From (7.6), we have

$$\operatorname{curl} \mathbf{A} = \left(-\frac{\partial}{\partial y} \int_0^x F_2\, dx - \frac{\partial}{\partial z} \int_0^x F_3\, dx + \xi \right) \mathbf{i} + F_2 \mathbf{j} + F_3 \mathbf{k}.$$

---

[1] Here there is an implicit assumption about the region $\mathscr{R}$, namely that axes $Oxyz$ can be so chosen that the straight lines RQ and QP lie entirely in $\mathscr{R}$. A proof for more general regions can be given, but the assumption made here covers most *singly connected* regions which are met in practice.

Differentiating through the integrals (which is permissible because the existence of div $\mathbf{F}$ implies that the functions F_1, F_2, F_3 are continuously differentiable), we obtain

$$\operatorname{curl} \mathbf{A} = -\left\{ \int_0^x \left(\frac{\partial F_2}{\partial y} + \frac{\partial F_3}{\partial z} \right) dx - \xi \right\} \mathbf{i} + F_2 \mathbf{j} + F_3 \mathbf{k}. \qquad (7.7)$$

But div $\mathbf{F} \equiv 0$, so that

$$\frac{\partial F_2}{\partial y} + \frac{\partial F_3}{\partial z} = -\frac{\partial F_1}{\partial x}.$$

Substituting this into (7.7) and integrating gives

$$\begin{aligned}
\operatorname{curl} \mathbf{A} &= ([F_1]_0^x + \xi)\mathbf{i} + F_2 \mathbf{j} + F_3 \mathbf{k} \\
&= F_1 \mathbf{i} + F_2 \mathbf{j} + F_3 \mathbf{k} = \mathbf{F},
\end{aligned}$$

as required. Thus (7.6) is a particular vector potential.

Uniqueness. Suppose that two vector fields $\mathbf{A}$ and $\mathbf{A}'$ are such that, in a singly connected region $\mathscr{R}$,

$$\mathbf{F} = \operatorname{curl} \mathbf{A} \quad \text{and} \quad \mathbf{F} = \operatorname{curl} \mathbf{A}'.$$

Write

$$\mathbf{B} = \mathbf{A} - \mathbf{A}'.$$

Then, by subtraction,

$$\operatorname{curl} \mathbf{B} \equiv \mathbf{0} \text{ in } \mathscr{R}.$$

It follows from the preceding section that there exists a scalar field Ω such that

$$\mathbf{B} = \operatorname{grad} \Omega.$$

Thus

$$\mathbf{A} = \mathbf{A}' + \operatorname{grad} \Omega, \qquad (7.8)$$

showing that *apart from the addition of the gradient of an arbitrary scalar field, the vector potential $\mathbf{A}$ is unique.*

Comment. From (7.8),

$$\operatorname{div} \mathbf{A} = \operatorname{div} \mathbf{A}' + \nabla^2 \Omega,$$

and so we have some latitude in the choice of div $\mathbf{A}$. In electromagnetism, this fact is exploited to simplify the form of the basic equations. For example, in electrostatics, it is usual to specify that div $\mathbf{A} \equiv 0$.

Solenoidal vector fields. A vector field $\mathbf{F}$ is said to be *solenoidal* in a region $\mathscr{R}$ if div $\mathbf{F} \equiv 0$ in $\mathscr{R}$.

EXERCISES

6. By using formula (7.6), find a vector field $\mathbf{A}$ such that $\mathbf{F} = \mathrm{curl}\,\mathbf{A}$ when $\mathbf{F} = (y - x, z, x)$.

7. Using spherical polar coordinates r, θ, ϕ, verify that the vector field $\mathbf{F} = \mathbf{e}_r/r^2$ is solenoidal. Find the function $\psi(r, \theta)$ such that

$$\mathbf{A} = \frac{\psi(r, \theta)}{r \sin \theta} \mathbf{e}_\phi$$

is the vector potential of $\mathbf{F}$, satisfying the condition $\mathrm{div}\,\mathbf{A} \equiv 0$.

7.4 Poisson's equation

In many applications, the scalar potential Ω satisfies the partial differential equation

$$\nabla^2 \Omega = f(x, y, z), \tag{7.9}$$

where $f(x, y, z)$ is a given bounded scalar function of position. This is known as *Poisson's equation*, except in the particular case $f(x, y, z) \equiv 0$, when it reduces to *Laplace's equation*.

The solution may be needed in a region τ bounded by a simple closed surface S for various *boundary conditions*. (For simplicity, we suppose that there is only one bounding surface, but the argument could easily be extended to the general case.) Typical boundary conditions are:

(i) the *Dirichlet condition*

$$\Omega = g(x, y, z) \quad \text{on } S, \tag{7.10}$$

where $g(x, y, z)$ is a given function of position;

(ii) the *Neumann condition*

$$\frac{\partial \Omega}{\partial n} = h(x, y, z) \quad \text{on } S, \tag{7.11}$$

where $h(x, y, z)$ is a given function of position, and $\partial \Omega / \partial n = \hat{\mathbf{n}} \cdot \nabla \Omega$, where $\hat{\mathbf{n}}$ denotes the unit normal on S.

Uniqueness. Generally, the solution of Poisson's equation is uniquely determined by either condition (7.10) or condition (7.11). This may be proved by supposing that Ω' is a second solution of (7.9), satisfying the same boundary conditions as Ω. Then writing

$$U = \Omega - \Omega'$$

we find that

$$\nabla^2 \Omega = f \quad \text{and} \quad \nabla^2 \Omega' = f,$$

giving by subtraction

$$\nabla^2 U = 0.$$

The boundary condition (7.10) gives

$$\Omega = g \quad \text{and} \quad \Omega' = g \quad \text{on } S;$$

therefore

$$U = 0 \quad \text{on } S.$$

It follows from the uniqueness theorem of § 6.3 that, throughout τ,

$$U \equiv 0.$$

Thus, the solution of Poisson's equation is uniquely determined by the boundary condition (7.10).

Similarly, it may be shown (the proof is set as an exercise) that the solution of Poisson's equation is unique, apart from an arbitrary additive constant, under a Neumann boundary condition.

It follows that the solution of the equation with *two* boundary conditions is possible only in very special circumstances. Problems with two boundary conditions are said to be *overdetermined*. In the cases when a solution exists satisfying both boundary conditions, the boundary conditions are said to be *compatible* or *dependent*.

Solution of Poisson's equation. Earlier (§ 6.3) we proved Green's theorem that for the scalar fields Φ and Ψ,

$$\int_S \left(\Phi \frac{\partial \Psi}{\partial n} - \Psi \frac{\partial \Phi}{\partial n} \right) dS = \int_\tau (\Phi \nabla^2 \Psi - \Psi \nabla^2 \Phi) \, d\tau. \tag{7.12}$$

Choose the origin O inside τ, and take

$$\Phi = \Omega, \quad \Psi = \frac{1}{r}. \tag{7.13}$$

Then, using the formula for ∇^2 in spherical polar coordinates,

$$\nabla^2 \Psi \equiv \frac{1}{r^2} \frac{d}{dr} \left\{ r^2 \frac{d}{dr} \left(\frac{1}{r} \right) \right\} \equiv 0 \tag{7.14}$$

except at the origin ($r = 0$). As $\nabla^2 \Psi$ is not defined at $r = 0$, we exclude this point and apply Green's theorem to the region $\tau - \tau_\epsilon$ bounded by $S \cup S_\epsilon$, as in Fig. 86, where S_ϵ denotes the surface of a small sphere of radius ϵ surrounding O, and τ_ϵ denotes the region bounded by S_ϵ. Substituting (7.13) and (7.14) into (7.12) gives

$$\int_{S \cup S_\epsilon} \left\{ \frac{1}{r} \frac{\partial \Omega}{\partial n} - \Omega \frac{\partial}{\partial n} \left(\frac{1}{r} \right) \right\} dS = \int_{\tau - \tau_\epsilon} \frac{1}{r} \nabla^2 \Omega \, d\tau;$$

that is

$$\int_S \left\{ \frac{1}{r}\frac{\partial \Omega}{\partial n} - \Omega \frac{\partial}{\partial n}\left(\frac{1}{r}\right) \right\} dS + I_\epsilon = \int_\tau \frac{1}{r}\nabla^2\Omega \, d\tau - J_\epsilon, \qquad (7.15)$$

where

$$I_\epsilon = \int_{S_\epsilon} \left\{ \frac{1}{r}\frac{\partial \Omega}{\partial n} - \Omega \frac{\partial}{\partial n}\left(\frac{1}{r}\right) \right\} dS \qquad (7.16)$$

and

$$J_\epsilon = \int_{\tau_\epsilon} \frac{1}{r}\nabla^2\Omega \, d\tau. \qquad (7.17)$$

We shall find the limiting form of (7.15) when $\epsilon \to 0$.

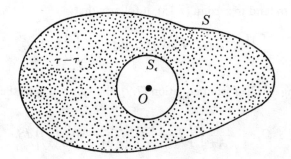

FIG. 86. S_ϵ is a small sphere, centre O; and $\tau - \tau_\epsilon$ is the region between S_ϵ and the surface S

By convention, the surface normal points out of $\tau - \tau_\epsilon$, and hence, on the boundary S_ϵ, $\partial/\partial n \equiv -\partial/\partial r$. Thus using spherical polar coordinates r, θ, ϕ we obtain

$$I_\epsilon = \int_0^{2\pi}\int_0^\pi \left(-\frac{1}{\epsilon}\frac{\partial \Omega}{\partial r} - \frac{\Omega}{\epsilon^2} \right) \epsilon^2 \sin\theta \, d\theta \, d\phi$$

$$= -\epsilon \int_0^{2\pi}\int_0^\pi \frac{\partial \Omega}{\partial r} \sin\theta \, d\theta \, d\phi - \int_0^{2\pi}\int_0^\pi \Omega(\epsilon, \theta, \phi) \sin\theta \, d\theta \, d\phi.$$

Hence, denoting by Ω_o the value of Ω at the origin O and noting that Ω is assumed to be continuous at this point,

$$\lim_{\epsilon \to 0} I_\epsilon = -\Omega_o \int_0^{2\pi}\int_0^\pi \sin\theta \, d\theta \, d\phi = -4\pi\Omega_o.$$

Also,

$$J_\epsilon = \int_0^{2\pi} \int_0^\pi \int_0^\epsilon r \nabla^2 \Omega \sin \theta \, dr \, d\theta \, d\phi,$$

and if M denotes the upper bound of $|\nabla^2 \Omega|$ (f is bounded in equation (7.9)), we have

$$|J_\epsilon| \leq M \left| \int_0^{2\pi} \int_0^\pi \int_0^\epsilon r \sin \theta \, dr \, d\theta \, d\phi \right|$$

$$= M 2\pi\epsilon^2.$$

Thus,

$$\lim_{\epsilon \to 0} J_\epsilon = 0.$$

Allowing ϵ to tend to zero in (7.15), it follows that

$$\int_S \left\{ \frac{1}{r} \frac{\partial \Omega}{\partial n} - \Omega \frac{\partial}{\partial n} \left(\frac{1}{r} \right) \right\} dS - 4\pi\Omega_o = \int_\tau \frac{1}{r} \nabla^2 \Omega \, d\tau. \tag{7.18}$$

Substituting $\nabla^2 \Omega = f$ from equation (7.9), the value of Ω at the origin O is thus

$$\Omega_o = -\frac{1}{4\pi} \int_\tau \frac{f}{r} d\tau + \frac{1}{4\pi} \int_S \left\{ \frac{1}{r} \frac{\partial \Omega}{\partial n} - \Omega \frac{\partial}{\partial n} \left(\frac{1}{r} \right) \right\} dS. \tag{7.19}$$

By choosing the origin O to be any field point, (7.19) gives the value of Ω at that point. However, this result does not give an immediate solution of Poisson's equation, because the right-hand side of (7.19) requires prior knowledge of Ω and $\partial\Omega/\partial n$ on S, and we have already seen that only one of these quantities is usually given there. The result is nevertheless useful because analytical and numerical methods are available for solving some equations of this type (called *integral equations*).

In the particular case when τ is a region of infinite extent and $\Omega = O(1/r)$ at large distances (in physical applications this condition is nearly always satisfied), expression (7.19) leads at once to the solution for Ω. For, take S to be a sphere of radius R centred at the origin. Then

$$\int_S \left\{ \frac{1}{r} \frac{\partial \Omega}{\partial n} - \Omega \frac{\partial}{\partial n} \left(\frac{1}{r} \right) \right\} dS = \int_0^{2\pi} \int_0^\pi \left(\frac{1}{r} \frac{\partial \Omega}{\partial r} + \frac{\Omega}{r^2} \right)_{r=R} R^2 \sin \theta \, d\theta \, d\phi$$

$$= \int_0^{2\pi} \int_0^\pi O\left(\frac{1}{R} \right) \sin \theta \, d\theta \, d\phi,$$

at large distances R. Thus, allowing R to tend to infinity, the surface integral vanishes, and (7.19) becomes

$$\Omega_o = -\frac{1}{4\pi} \int \frac{f}{r} d\tau, \qquad (7.20)$$

where the integration is now taken over all space and

$$r = (x^2 + y^2 + z^2)^{\frac{1}{2}}. \qquad (7.21)$$

Changing the origin to the point (x, y, z), we obtain the alternative form

$$\Omega(x, y, z) = -\frac{1}{4\pi} \int\int\int \frac{f(x', y', z')}{\{(x-x')^2 + (y-y')^2 + (z-z')^2\}^{\frac{1}{2}}} dx' \, dy' \, dz'. \quad (7.22)$$

This is the required solution of Poisson's equation.

EXERCISES

8. Prove that the solution of Poisson's equation under a Neumann boundary condition is unique, apart from an arbitrary additive constant.

9. The solution of Poisson's equation $\nabla^2\Omega = f(x, y, z)$, where $f(x, y, z)$ is a given scalar field, is required in a region τ bounded by a simple closed surface S. The boundary condition is

$$\Omega = g(x, y, z) \quad \text{on } S,$$

where $g(x, y, z)$ is a given scalar function of position. By taking $f(x, y, z) \equiv 0$ outside S, reduce the problem to that of solving Laplace's equation $\nabla^2 U = 0$ under the boundary condition

$$U = g(x, y, z) + \frac{1}{4\pi} \int\int\int \frac{f(x', y', z')}{\sqrt{\{(x-x')^2 + (y-y')^2 + (z-z')^2\}}} dx' \, dy' \, dz' \quad \text{on } S,$$

where the integral is taken over all space.

10. In spherical polar coordinates r, θ, ϕ, when Ω is a function of r only, the equation $\nabla^2\Omega = 1$ becomes

$$\frac{1}{r^2} \frac{d}{dr}\left(r^2 \frac{d\Omega}{dr}\right) = 1.$$

Deduce that the continuous solution of the equation $\nabla^2\Omega = 1$ under the boundary condition $\Omega = 1$ on the sphere $r = 1$ is

$$\Omega = (5 + r^2)/6.$$

11. If $\nabla^2\Omega = 1$ for $r \leqslant 1$ and $\nabla^2\Omega = 0$ for $r > 1$, show that equation (7.22) yields the solution

$$\Omega = \tfrac{1}{6}r^2 - \tfrac{1}{2}, \quad r \leqslant 1.$$

Using the method of Exercise 9, obtain from this result the solution to Exercise 10.

7.5 Poisson's equation in vector form

In electromagnetism a vector potential $\mathbf{A}$ occurs and satisfies Poisson's equation in vector form, viz.

$$\nabla^2 \mathbf{A} = \mathbf{F}, \tag{7.23}$$

where $\mathbf{F}$ is a known vector field. If the rectangular cartesian components of $\mathbf{A}$ are A_1, A_2, A_3 and those of $\mathbf{F}$ are F_1, F_2, F_3, the components of (7.23) are

$$\nabla^2 A_i = F_i \quad (i = 1, 2, 3), \tag{7.24}$$

which is the scalar form of Poisson's equation.

If the region under consideration is unbounded, and $A_i = O(1/r)$ at large distance r from the origin, the solution of (7.24) is

$$A_i(x, y, z) = -\frac{1}{4\pi} \int\!\!\int\!\!\int \frac{F_i(x', y', z')}{\{(x-x')^2 + (y-y')^2 + (z-z')^2\}^{\frac{1}{2}}}\, dx'\, dy'\, dz'. \tag{7.25}$$

Combining the components to recover $\mathbf{A}$ gives

$$\mathbf{A}(x, y, z) = -\frac{1}{4\pi} \int\!\!\int\!\!\int \frac{\mathbf{F}(x', y', z')}{\{(x-x')^2 + (y-y')^2 + (z-z')^2\}^{\frac{1}{2}}}\, dx'\, dy'\, dz'. \tag{7.26}$$

7.6 Helmholtz's theorem

This theorem states that for any continuously differentiable vector field $\mathbf{H}$ there exists a scalar field Ω and a vector field $\mathbf{A}$ such that

$$\mathbf{H} = \operatorname{grad}\Omega + \operatorname{curl}\mathbf{A}. \tag{7.27}$$

Here we shall suppose that the region under consideration is unbounded, and that $|\mathbf{H}| = O(1/r^2)$ at large distances.

Proof. Consider the differential equation

$$\nabla^2 \Omega = \operatorname{div}\mathbf{H}$$
$$= f(x, y, z), \quad \text{say.} \tag{7.28}$$

Imposing the condition $\Omega = O(1/r)$ at large distances, an explicit formula for Ω is given by (7.22), namely,

$$\Omega(x, y, z) = -\frac{1}{4\pi} \int\!\!\int\!\!\int \frac{f(x', y', z')}{\{(x-x')^2 + (y-y')^2 + (z-z')^2\}^{\frac{1}{2}}}\, dx'\, dy'\, dz'. \tag{7.29}$$

Equation (7.28) can also be expressed as

$$\operatorname{div}(\mathbf{H} - \operatorname{grad}\Omega) = 0.$$

It follows from § 7.3, p. 190, that there exists a vector field $\mathbf{A}$ such that

$$\mathbf{H} - \operatorname{grad}\Omega = \operatorname{curl}\mathbf{A}.$$

This proves the theorem.

An explicit formula for **A** can be obtained as follows. Taking the curl of (7.27) gives

$$\text{curl curl }\mathbf{A} = \text{curl }\mathbf{H}. \tag{7.30}$$

As we remarked at the end of § 7.3, the value of div**A** can be chosen arbitrarily. We now use this freedom to impose the condition div**A** = 0. Then, as $\nabla^2\mathbf{A} \equiv \text{grad}(\text{div}\,\mathbf{A}) - \text{curl curl }\mathbf{A}$, (7.30) becomes

$$\begin{aligned} \nabla^2\mathbf{A} &= -\text{curl }\mathbf{H} \\ &= \mathbf{F}(x,y,z), \quad \text{say.} \end{aligned} \tag{7.31}$$

Using the solution (7.26), we thus have

$$\mathbf{A}(x,y,z) = -\frac{1}{4\pi}\iiint \frac{\mathbf{F}(x',y',z')}{\{(x-x')^2+(y-y')^2+(z-z')^2\}^{\frac{1}{2}}}\,dx'\,dy'\,dz', \tag{7.32}$$

which is the required explicit formula for **A**. That this result is consistent with the condition div**A** = 0 can be seen as follows.

Take the divergence of (7.31). Then

$$\nabla^2(\text{div}\,\mathbf{A}) = 0. \tag{7.33}$$

At large distances, **A** (as given by (7.32)) is seen to be $O(1/r)$, and so div**A** = $O(1/r^2)$. As there are no internal boundaries, a particular solution is evidently div**A** ≡ 0. By the uniqueness theorem proved in § 6.3, it follows that this is the only possible solution, and hence **A** as given by (7.32) must be such that div**A** = 0.

Comments. (1) The vector fields gradΩ and curl**A** in (7.27) are, respectively, irrotational and solenoidal; that is curl(gradΩ) ≡ **0** and div(curl**A**) ≡ 0. Helmoltz's theorem thus shows that the vector field **H** can be resolved into the sum of an irrotational part and a solenoidal part.

(2) Expressions (7.29) and (7.32) show how to determine Ω and **A** from div**H** and curl**H**. Regions where div**H** ≠ 0 are called *sources* of **H**; and regions where curl**H** ≠ **0** are called *vortices* of **H**. This terminology originates in hydrodynamics.

7.7 Solid angles

Finally, we give a brief account (mainly in geometrical terms) of solid angles, since these occur frequently in potential theory.

Let dS be an element of surface area at a point P of a surface S, and let O be any other point. Denote by θ the angle between $\overrightarrow{OP}$ and the unit normal **n** to S at P (Fig. 87). If $\overrightarrow{OP} = \mathbf{r}$, we define

$$d\omega = \frac{\cos\theta\,dS}{r^2} \tag{7.34}$$

to be the *solid angle* subtended by dS at O. The total solid angle subtended by S at O is

$$\omega = \int_S \frac{\cos\theta\, dS}{r^2} = \int_S \frac{\mathbf{r}.\mathbf{dS}}{r^3}. \tag{7.35}$$

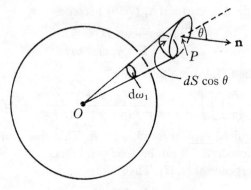

FIG. 87

Consider a cone with base dS and vertex O. Draw a sphere with centre O and of unit radius, and let $d\omega_1$ be the area of surface cut off by the cone (Fig. 87). The projection of the area dS onto the plane perpendicular to OP is $dS|\cos\theta|$. The ratio of this area to $d\omega_1$ is $OP^2:1$ and so

$$d\omega_1 = \frac{dS|\cos\theta|}{r^2} = |d\omega|. \tag{7.36}$$

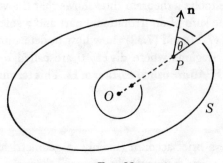

FIG. 88

Thus, if θ is an acute angle (so that $|\cos\theta| = \cos\theta$), the solid angle subtended by dS at O is just the area cut off by the cone on the sphere of unit radius and center at O; if θ is an obtuse angle, the solid angle is $(-1) \times$ area cut off. This intuitive geometrical argument often enables a solid angle to be determined more easily than by performing the integrations in (7.35).

Special Cases. (1) Consider a closed surface S which is such that all straight lines radiating from points inside it cut the surface once only (Fig. 88). If O is any point inside S and the unit normals **n** at points P on S are directed away from the interior (in the conventional way), the angles θ between $\overrightarrow{OP}$ and **n** are all acute. The solid angle subtended by S at O is therefore equal to the surface area of a sphere of unit radius, which is 4π.

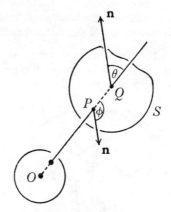

FIG. 89

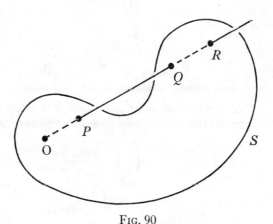

FIG. 90

Suppose next that O lies outside S, and let a line radiating from O cut S at P and Q (Fig. 89). The angles between OPQ and the unit normals at P and Q (ϕ and θ in the diagram) are respectively obtuse and acute. Draw a slender cone with vertex at O, cutting off elements of area at P and Q. The solid angles subtended at O by these elements are equal in magnitude but have opposite signs; their net contribution to the total solid angle subtended by S at O is therefore zero. This holds for all lines OPQ, so the solid angle subtended by S at O is zero.

If O lies on S, a similar argument may be used to show that the solid angle is 2π.

Summarizing, *the solid angle subtended by S at O is 4π, 2π or zero according as O lies inside, on or outside S, respectively.*

The same results hold for more general surfaces S, such as that shown in Fig. 90. When O lies inside S in the position indicated, and a line radiating from O cuts S at P, Q and R, the contributions to the solid angle from surface elements at P, Q, R are equal in magnitude, but two are positive and one is negative; hence the net contribution is the same as if $OPQR$ cut S only once. Similarly, any line $\mathscr{L}$ radiating from a point outside S cuts the surface an even number of times and so the total contribution to the solid angle from the surface elements on $\mathscr{L}$ is zero. The general result for any closed surface can therefore be established.

(2) Let O be a point on the axis of a circular disc, and distant a from it, and suppose the disc is oriented so that the unit normal $\mathbf{n}$ points away from O, as in Fig. 91. Let α be the semi-vertical angle of the cone which has the disc as base

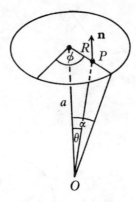

FIG. 91

and vertex at O. Taking cylindrical polar coordinates with origin at O and axis coinciding with the axis of the disc, the solid angle subtended at O is

$$\omega = \int_S \frac{\cos\theta \, dS}{r^2} = \int_{-\pi}^{\pi} \int_0^{a\tan\alpha} \frac{\cos\theta \, R \, dR \, d\phi}{a^2 \sec^2\theta}$$

$$= 2\pi \int_0^{a\tan\alpha} \cos^3\theta \, R \, dR / a^2.$$

Putting $R = a\tan\theta$, this becomes

$$\omega = 2\pi \int_0^a \sin\theta \, d\theta$$

$$= 2\pi(1 - \cos\alpha).$$

By taking $\alpha = \frac{1}{2}\pi$, it also follows from this that an infinite plane subtends a solid angle of magnitude 2π at any point which does not lie on the plane.

EXERCISES

12. By using the divergence theorem, prove that the solid angle subtended by a simple closed surface at an exterior point is zero.

13. Points P, Q are situated on opposite sides of a simple open surface S and are arbitrarily close to the surface. The solid angles subtended by S at P and Q are denoted by ω_P^S and ω_Q^S, respectively. If P is on the positive side of S, prove that

$$\omega_P^S - \omega_Q^S \to -4\pi \quad \text{as} \quad PQ \to 0.$$

[*Hint.* Join to S a simple open surface S', such that $S \cup S'$ is a simple closed surface surrounding Q; and observe that

$$\omega_P^{S'} \to \omega_Q^{S'} \quad \text{as} \quad PQ \to 0.]$$

14. A square has sides of length $2a$, and a point P is situated at a distance a from the square on a normal through the intersection of its diagonals. If the unit normal points away from P, find the solid angle subtended by the square at this point.

CHAPTER 8

CARTESIAN TENSORS

8.1 Introduction

We have seen that the essential feature of a vector is that, in any rectangular cartesian coordinate system, it may be represented by three components, each associated with a particular axis. These components depend only on the orientation of the axes and transform according to the rules (2.1) when the axes are rotated. Tensor analysis may be regarded as a generalisation of vector analysis to certain mathematical and physical entities known as tensors which require more than three components for their complete specification. There are again physically meaningful rules for transforming the components of tensors when the axes are changed. To give some motivation for the study of tensors, we shall first provide a specific example.

When an elastic body is deformed, certain internal forces will, in general, be set up. At any particular point P within such a body, consider a small plane area whose normal is in the direction of the x_1-axis of a rectangular cartesian system $Ox_1x_2x_3$. The force per unit area on this element due to the material in contact with it is a vector having three rectangular cartesian components, which may be denoted by σ_{11}, σ_{12} and σ_{13}. Similarly, the forces per unit area on small plane areas at P whose normals are in the directions Ox_2 and Ox_3, respectively, will have components which may be denoted by σ_{21}, σ_{22}, σ_{23} and σ_{31}, σ_{32}, σ_{33}, respectively. It may be shown that the nine components σ_{ij} $(i, j = 1, 2, 3)$ are sufficient to specify fully the 'state of stress' at the point P. Further, by applying Newton's laws of motion to an elementary part of the body, it is possible to deduce just how these components transform under a rotation of the axes.

There are many other mathematical and physical entities which may be represented in a similar way to stress, and this provides the motivation for setting up a comprehensive and general theory.

The purpose of this chapter is to give an introduction to *cartesian tensors*, that is tensors which are expressed in terms of components referred to rectangular cartesian coordinate systems. However, it is worth noting that, as with vectors, orthogonal curvilinear coordinates can often be handled with this framework. The remark after Example 1 later in this chapter (p. 214) illustrates this. General tensor analysis, which is an indispensable tool in the general theory of relativity and the study of non-Euclidean geometries, uses general coordinates; the concept of space itself is generalized. It is therefore, a much more sophisticated theory and falls well outside the intended scope of this volume.

8.2   Cartesian tensors: basic algebra

Throughout the remainder of this chapter, $Ox_1x_2x_3$ and $Ox_1'x_2'x_3'$ denote rectangular cartesian coordinate systems whose relative orientations are given, as in Chapter 1, by the transformation matrix

$$
\begin{array}{c|ccc}
 & x_1 & x_2 & x_3 \\
\hline
x_1' & l_{11} & l_{12} & l_{13} \\
x_2' & l_{21} & l_{22} & l_{23} \\
x_3' & l_{31} & l_{32} & l_{33}
\end{array}
\tag{8.1}
$$

The summation convention will also be used throughout.

DEFINITION. Suppose that A is a mathematical or physical entity which, when it is associated with a rectangular cartesian coordinate system $Ox_1x_2x_3$, may be represented by a set of 3^n scalars $a_{ij...}$, where there are $n(\geqslant 0)$ suffixes attached to a. Each suffix takes one of the values 1, 2 or 3 and the suffixes $i, j, \ldots$ are ordered. Let $a_{rs...}'$ (n suffixes) be the corresponding set of scalars with ordered suffixes representing A when it is referred to the axes $Ox_1'x_2'x_3'$, and suppose that

$$
a_{rs...}' = l_{ri}l_{sj}\cdots a_{ij...}.
\tag{8.2}
$$

If the sets of scalars with ordered suffixes representing A are also invariant under a translation of the axes (and hence depend only on the orientation of the axes), A is said to be a *tensor of order (or rank)* n; the scalars $a_{ij...}$ are called components of A.

A tensor of order zero is evidently a single scalar a having the same value in all coordinate systems; it is therefore a *scalar invariant*.

In the particular case when $n = 1$, A has three components, a_1, a_2 and a_3, and the transformation (8.2) reduces to

$$
a_r' = l_{ri}a_i.
\tag{8.3}
$$

This is the same as the *vector* transformation law, and so a *tensor of order one is a vector*. This confirms our earlier statement that cartesian tensor analysis may be regarded as a generalisation of vector analysis.

When $n = 2$, the tensor A is of second order and has nine components which may be conveniently assembled to form the matrix

$$\begin{pmatrix} a_{11} & a_{12} & a_{13} \\ a_{21} & a_{22} & a_{23} \\ a_{31} & a_{32} & a_{33} \end{pmatrix} \tag{8.4}$$

It should be noted that the elements of this matrix may change from one set of axes to another. The transformation rule (8.2) becomes

$$a'_{rs} = l_{ri} l_{sj} a_{ij}. \tag{8.5}$$

Observe that the components are ordered in the sense that a_{ij} is associated with Ox_i and Ox_j in that order; a_{11} is therefore associated with Ox_1 twice, a_{12} with Ox_1 and Ox_2 in that order, a_{21} with Ox_2 and Ox_1 in that order, and so on. In the particular case of the stress tensor with components σ_{ij}, mentioned in §8.1, the first suffix indicates the direction of the normal to an element under consideration and the second suffix the direction of the component of the stress acting on that element.

Apart from scalar invariants and vectors, the most commonly occurring tensors are those of second order; the main emphasis in the remainder of this chapter will be on tensors of this type.

Elementary algebraic operations. The definitions of multiplication of a tensor by a scalar, and addition and subtraction of tensors *of the same order* are analogous to the corresponding definitions of these operations in §'s 2.3 and 2.4 for vectors. For example, if A is a second order tensor with components a_{ij} $(i, j = 1, 2, 3)$ and λ is a scalar, then λA is a tensor with components λa_{ij}. Further, if B is another second order tensor with components b_{ij}, then $A + B$ is a tensor with components $a_{ij} + b_{ij}$. Addition (and subtraction) of tensors of different orders is not defined.

It may be readily proved from the definition of a tensor that the operation of addition is commutative and associative (cf. § 2.4).

Product of tensors. Let A and B be tensors of orders m and n, respectively, whose components referred to the axes $Ox_1 x_2 x_3$ are $a_{ij...}$ and $b_{pq...}$, there being m suffixes attached to a and n suffixes attached to b. Then the 3^{m+n} scalars

$$c_{ij...pq...} = a_{ij...} b_{pq...} \tag{8.6}$$

form a tensor C, say, of order $m + n$. This is defined to be the product of A and B and is written as $C = AB$.

Proof. Under the rotation of axes from the set $Ox_1x_2x_3$ to the set $Ox_1'x_2'x_3'$, defined by the transformation matrix (8.1), the transformed components of A and B are

$$a'_{rs...} = l_{ri}l_{sj}\cdots a_{ij...}$$

and

$$b'_{uv...} = l_{up}l_{vq}\cdots b_{pq...},$$

respectively. Hence, relative to the new axes, the components of C are

$$\begin{aligned}c'_{rs...uv...} &= a'_{rs...}b'_{uv...} \\ &= l_{ri}l_{sj}\cdots l_{up}l_{vq}\cdots a_{ij...}b_{pq...} \\ &= l_{ri}l_{sj}\cdots l_{up}l_{vq}\cdots c_{ij...pq...}.\end{aligned}$$

This is the tensor law of transformation, and the tensorial character of C is therefore established.

As a particular example of this important operation, consider the vectors $\mathbf{a} = (a_1, a_2, a_3)$ and $\mathbf{b} = (b_1, b_2, b_3)$. As said before, these are first order tensors, and by the result just proved their product is a second order tensor with 9 components.

$$c_{ij} = a_i b_j \quad (i, j = 1, 2, 3).$$

The operation of tensor multiplication is associative, i.e.

$$(AB)C = A(BC). \tag{8.7}$$

The distributive law

$$A(B+C) = AB + AC \tag{8.8}$$

is also obeyed. The proofs of these results are left for the reader as exercises.

Contraction. The operation of making two suffixes in the components of a tensor equal and then summing over the repeated suffix is called *contraction*. If a tensor is of order $n(\geqslant 2)$, contraction over any pair of suffixes gives a tensor of order $n-2$. The proof of this is left as an exercise.

As an example, consider the second order tensor with components $a_i b_j$, which is the (tensor) product of the vectors $\mathbf{a} = (a_1, a_2, a_3)$ and $\mathbf{b} = (b_1, b_2, b_3)$. Contracting over the suffixes i and j gives

$$a_i b_i = a_1 b_1 + a_2 b_2 + a_3 b_3,$$

which is a tensor of order zero (a scalar invariant). It is, of course, the scalar product of $\mathbf{a}$ and $\mathbf{b}$.

Symmetry and anti-symmetry. If a tensor A with components a_{ij} is such that

$$a_{ij} = a_{ji}$$

for all i and j, then A is said to be *symmetrical*. If

$$a_{ij} = -a_{ji}$$

for all i and j, then A is said to be *anti-symmetrical*. Observe that when A is anti-symmetrical, the components a_{11}, a_{22} and a_{33} must be zero.

This definition may be extended to tensors of higher order, the symmetry or anti-symmetry being defined with respect to a particular pair of suffixes.

Every second order tensor can be represented as the sum of a symmetrical and an anti-symmetrical tensor. For, if a tensor has components a_{ij}, then

$$a_{ij} = \tfrac{1}{2}(a_{ij} + a_{ji}) + \tfrac{1}{2}(a_{ij} - a_{ji}),$$

and it is readily seen that the two expressions in parentheses on the right-hand side are, respectively, the components of symmetrical and anti-symmetrical tensors.

Symmetry and anti-symmetry are *intrinsic* properties of a tensor, being independent of the coordinate system in which it is represented. To prove this, let A have components a_{ij} relative to axes $Ox_1x_2x_3$. Then, relative to the axes $Ox_1'x_2'x_3'$,

$$\begin{aligned}
a_{rs}' &= l_{ri}l_{sj}a_{ij} \\
&= l_{rj}l_{si}a_{ji},
\end{aligned}$$

where in the last line, the dummy suffixes i and j over which the expression is summed have been interchanged. If A is symmetrical with respect to the axes $Ox_1x_2x_3$, $a_{ji} = a_{ij}$. It follows that

$$a_{rs}' = l_{si}l_{rj}a_{ij} = a_{sr}',$$

which shows that A is also symmetrical with respect to the axes $Ox_1'x_2'x_3'$. Similarly, the property of anti-symmetry is preserved when the axes are changed.

The quotient rule. Let A be a mathematical or physical entity which, when it is associated with any set of rectangular cartesian coordinate axes $Ox_1x_2x_3$, may be represented by an ordered set of nine scalars a_{ij} ($i, j = 1, 2, 3$). Suppose that for all vectors $\mathbf{v} = (v_1, v_2, v_3)$, the scalars

$$u_i = a_{ij}v_j \qquad (i = 1, 2, 3) \tag{8.9}$$

are the components of a vector $\mathbf{u}$. Then A is a second order tensor.

Proof. Under a translation of the axes let the components a_{ij} transform to a'_{ij}. Since the components of **u** and **v** are invariant under such a translation

$$u_i = a'_{ij} v_j.$$

Subtracting this from (8.9) gives

$$(a_{ij} - a'_{ij}) v_j = 0,$$

and since this holds for all vectors **v**, it follows that $a'_{ij} = a_{ij}$. The components of A are therefore invariant under a translation of the axes.

Consider next a rotation to new axes $Ox'_1 x'_2 x'_3$ and let the components of **u**, **v** and A transform to u'_i, v'_i and a'_{ij}, respectively. Then

$$\begin{aligned} a'_{r\alpha} v'_\alpha &= u'_r \qquad (r = 1, 2, 3) \\ &= l_{ri} u_i \\ &= l_{ri} a_{ij} v_j. \end{aligned}$$

But

$$v'_\alpha = l_{\alpha j} v_j,$$

and hence $(a'_{r\alpha} l_{\alpha j} - l_{ri} a_{ij}) v_j = 0$.

As this holds for all vectors **v**, the coefficients of v_1, v_2 and v_3 must vanish and hence

$$a'_{r\alpha} l_{\alpha j} = l_{ri} a_{ij}.$$

Multiplying throughout by l_{sj} and summing over the suffix j gives

$$a'_{r\alpha} l_{\alpha j} l_{sj} = l_{ri} l_{sj} a_{ij}.$$

But

$$l_{\alpha j} l_{sj} = \delta_{\alpha s},$$

and hence

$$a'_{rs} = l_{ri} l_{sj} a_{ij}.$$

The components of A therefore satisfy the tensor laws under a translation and rotation of the axes, which proves that A is a second order tensor.

The result just proved extends to tensors of higher order and is usually called the *quotient rule*. In the general case, (8.9) is replaced by

$$u_{ij\ldots} = a_{ij\ldots rs\ldots} v_{rs\ldots}, \qquad (8.10)$$

where $u_{ij\ldots}$ and $v_{rs\ldots}$ are components of tensors U and V of orders m and n, respectively, and the inference which may be drawn is that, if (8.10) holds for all tensors V of order n, then $a_{ij\ldots rs\ldots}$ are the components of a tensor of order $m + n$. The proof is similar to that above.

EXERCISES

1. Relative to rectangular cartesian coordinate axes $Ox_1x_2x_3$, the second order tensor A has components a_{ij}. Another rectangular cartesian coordinate system $Ox_1'x_2'x_3'$ is chosen such that Ox_1' coincides with Ox_2 and Ox_3' coincides with Ox_3. Find the components of A relative to the new axes, and verify that the tensor remains symmetrical if it is so initially.

2. Show from first principles that the scalar obtained by contracting any second order tensor over its two suffixes is invariant under a rotation of the axes [*Hint.* Use the orthonormality conditions $l_{ri}l_{rj} = \delta_{ij}$].

3. The tensors A and B with components a_{ij} and b_{ij} are given to be symmetrical and anti-symmetrical respectively. Show that $a_{ij}b_{ij} = 0$.

4. The set of scalars u_{ijkm} is such that for all second order tensors with components E_{km}, the scalars

$$\sigma_{ij} = u_{ijkm}E_{km} \qquad (i, j = 1, 2, 3)$$

are the components of a second order tensor. Prove that the scalars u_{ijkm} are the components of a 4th order tensor.

[*Hint.* This is a particular case of the quotient rule (8.10) and may be proved by a method similar to that used to prove the special case (8.9). To establish that the components u_{ijkm}' transform correctly under a rotation of the axes, begin by considering $u_{rs\alpha\beta}'E_{\alpha\beta}' = \sigma_{rs}'$.]

8.3 Isotropic tensors

A cartesian tensor is said to be *isotropic* if its components are identical in all rectangular cartesian coordinate systems. Such tensors play a fundamental role in theoretical studies of materials whose physical properties are independent of direction. This section is concerned with the isotropic tensors of second, third and fourth orders, which are of particular importance.

The delta tensor. In the second order tensor law

$$a_{rs}' = l_{ri}l_{sj}a_{ij},$$

set $a_{ij} = \delta_{ij}$ where δ_{ij} is the Kronecker delta defined in §1.6. Then using the orthonormality conditions (1.28),

$$a_{rs}' = l_{ri}l_{si} = \delta_{rs}.$$

This shows that when the numbers δ_{11}, δ_{12}, ... are associated with a set of rectangular cartesian coordinate axes, they transform into themselves under the tensor rotation law. If the association with the axes is made independent of the choice of origin, so that invariance under a translation of the axes is assured, the numbers δ_{ij} evidently form a second order isotropic tensor. It is called the *delta tensor* and is the most important of all the isotropic tensors.

Even and odd permutations. Let i, j, k be a permutation of the numbers 1, 2, 3. The permutation is said to be *even* if i, j, k are three consecutive numbers of the set 1, 2, 3, 1, 2, and is said to be *odd* otherwise. For example, 2, 3, 1 is an even permutation of the numbers 1, 2, 3, and 2, 1, 3 is an odd permutation. It may be readily verified that all even permutations of the numbers 1, 2, 3 can be brought about by an even number of interchanges of pairs of these numbers and all odd permutations by an odd number of interchanges.

The alternating tensor. Suppose that each of the suffixes i, j and k in the symbol ϵ_{ijk} can take any one of the values 1, 2 or 3. Let

$$\epsilon_{ijk} = \begin{cases} 0 \text{ if two or more of the suffixes } i, j, k \text{ are equal} \\ 1 \text{ if } i, j, k \text{ is an even permutation of the numbers 1, 2, 3} \\ -1 \text{ if } i, j, k \text{ is an odd permutation of the numbers 1, 2, 3} \end{cases}$$

$$(8.11)$$

The symbol ϵ_{ijk} is called the *alternator*.

Set $a_{ijk} = \epsilon_{ijk}$ in the third order tensor rotation law

$$a'_{rst} = l_{ri}l_{sj}l_{tk}a_{ijk}$$

Then

$$a'_{rst} = l_{ri}l_{sj}l_{tk}\epsilon_{ijk} = l_{r1}l_{s2}l_{t3} + l_{r2}l_{s3}l_{t1} + l_{r3}l_{s1}l_{t2}$$
$$- l_{r1}l_{s3}l_{t2} - l_{r2}l_{s1}l_{t3} - l_{r3}l_{s2}l_{t1}$$

$$= \begin{vmatrix} l_{r1} & l_{s1} & l_{t1} \\ l_{r2} & l_{s2} & l_{t2} \\ l_{r3} & l_{s3} & l_{t3} \end{vmatrix}; \qquad (8.12)$$

the latter expression follows from the definition of a third order determinant, give in Appendix 1.

If any pair of the suffixes r, s, t are equal, the determinant (8.12) has two equal columns and hence vanishes. If $r = 1$, $s = 2$ and $t = 3$, the determinant reduces to the transpose of that given in equation (1.17) and it then has the value 1. Interchanging any pair of columns in a determinant changes its sign (Appendix 1), and hence if $r = 2$, $s = 1$ and $t = 3$, the determinant has the value -1. Now, all even permutations of the numbers 1, 2, 3 can be brought about by an even number of interchanges of pairs of these numbers, and all permutations which are odd can be brought about by an even number of interchanges of pairs of the numbers 2, 1, 3. The sign (and value) of a determinant is left unchanged by an even number of interchanges of pairs of its columns, and it thus follows that, for all possible values of r, s and t the determinant (8.12) has the same value as ϵ_{rst}. Hence

$$a'_{rst} = \epsilon_{rst},$$

which shows that, under a rotation of the axes, the tensor law is satisfied and each one of the set of numbers ϵ_{ijk} transforms into itself. If the association with the axes is defined to be independent of the choice of origin, then ϵ_{ijk} are the components of a third order isotropic tensor called the *alternating tensor*.

Relation between the alternating tensor and delta tensor. A very useful and important relation between the alternating and delta tensors is that

$$\epsilon_{ijk}\epsilon_{rsk} = \delta_{ir}\,\delta_{js} - \delta_{is}\,\delta_{jr}. \tag{8.13}$$

This may be established as follows.

If $i = j$ or $r = s$, the right-hand side of (8.13) is zero and the left-hand side also vanishes by the definition of the alternator.

Consider the case when $i \neq j$ and $r \neq s$. Without loss of generality we may choose $i = 1$ and $j = 2$. Using the definition of the alternator, the left-hand side of (8.13) then becomes

$$\epsilon_{121}\epsilon_{rs1} + \epsilon_{122}\epsilon_{rs2} + \epsilon_{123}\epsilon_{rs3} = \epsilon_{rs3}.$$

The right-hand side of (8.13) becomes

$$\delta_{1r}\,\delta_{2s} - \delta_{1s}\,\delta_{2r} = \Delta, \text{ say.}$$

As $r \neq s$, there are just the following possibilities to consider:

$r = 3$ in which case $\Delta = 0$ for all s;
$s = 3$ in which case $\Delta = 0$ for all r;
$r = 1$, $s = 2$, giving $\Delta = 1$;
$r = 2$, $s = 1$ giving $\Delta = -1$.

Hence $\Delta = \epsilon_{rs3}$, and the identity is proved.

Another way of proving the identity is indicated in Exercise (13), p. 218.

Relation of the alternating tensor to the vector product. The i-th component of the vector product of $\mathbf{a} = (a_1, a_2, a_3)$ with $\mathbf{b} = (b_1, b_2, b_3)$ is related to ϵ_{ijk} by the formula

$$(\mathbf{a} \times \mathbf{b})_i = \epsilon_{ijk}a_j b_k. \tag{8.14}$$

To prove this, consider the x_1-component. When $i = 1$, ϵ_{ijk} is non-zero only when $j = 2$ and $k = 3$ or when $j = 3$ and $k = 2$. Hence, according to (8.14),

$$(\mathbf{a} \times \mathbf{b})_1 = \epsilon_{123} a_2 b_3 + \epsilon_{132} a_3 b_2 = a_2 b_3 - a_3 b_2,$$

which verifies that the fomula gives the correct x_1-component. Likewise the x_2- and x_3-components are correctly represented.

Observe that $\epsilon_{ijk} a_j b_k$ is the product $\epsilon_{ijk} a_r b_s$ contracted over the suffixes j, r and k, s. Since the alternating tensor is of third order and $\mathbf{a}$ and $\mathbf{b}$ are first order tensors, their product is a fifth order tensor. Contraction of this over *two* pairs of suffixes reduces the order by 4 (see §8.2) yielding a first order tensor which is the vector $\mathbf{a} \times \mathbf{b}$. This provides an alternative to the proof in § 2.7 that $\mathbf{a} \times \mathbf{b}$ is a vector.

Another formula, closely associated with (8.14) is that for any vector field $\mathbf{F}$,

$$(\text{curl } \mathbf{F})_i = \epsilon_{ijk} \, \partial F_k / \partial x_j. \tag{8.15}$$

The reader should have no difficulty in proving this.

EXAMPLE 1. In an incompressible fluid, the stress on an element whose normal is in the direction of a unit vector $\mathbf{n}$ has components $t_i(\mathbf{n}) = \sigma_{ij} n_j$ ($i = 1, 2, 3$), where σ_{ij} is the stress tensor. Given that

$$\sigma_{ij} = -p \delta_{ij} + \mu \left(\frac{\partial v_i}{\partial x_j} + \frac{\partial v_j}{\partial x_i} \right), \tag{8.16}$$

verify that

$$\mathbf{t}(\mathbf{n}) = -p\mathbf{n} + 2\mu(\mathbf{n} \cdot \nabla)\mathbf{v} + \mu \mathbf{n} \times \text{curl } \mathbf{v}. \tag{8.17}$$

Solution. Using (8.14), the i-th component of $\mathbf{n} \times \text{curl } \mathbf{v}$ is

$$(\mathbf{n} \times \text{curl } \mathbf{v})_i = \epsilon_{ijk} n_j (\text{curl } \mathbf{v})_k$$
$$= \epsilon_{ijk} n_j \epsilon_{krs} \partial v_s / \partial x_r,$$

by (8.15). But

$$\epsilon_{krs} = -\epsilon_{rks} = \epsilon_{rsk},$$

because interchanging any pair of suffixes in the alternator changes its sign. Hence,

$$(\mathbf{n} \times \text{curl } \mathbf{v})_i = \epsilon_{ijk} \epsilon_{rsk} n_j \partial v_s / \partial x_r$$
$$= (\delta_{ir} \delta_{js} - \delta_{is} \delta_{jr}) n_j \partial v_s / \partial x_r$$
$$= n_j \frac{\partial v_j}{\partial x_i} - n_j \frac{\partial v_i}{\partial x_j}.$$

The i-th component of (8.17) is therefore

$$t_i(\mathbf{n}) = -pn_i + 2\mu n_j \frac{\partial v_i}{\partial x_j} + \mu(\mathbf{n} \times \text{curl } \mathbf{v})_i$$

$$= -pn_i + \mu n_j \left(\frac{\partial v_i}{\partial x_j} + \frac{\partial v_j}{\partial x_i} \right). \tag{8.18}$$

But, we are given that

$$t_i(\mathbf{n}) = \sigma_{ij} n_j,$$

and substituting σ_{ij} from (8.16) an expression identical to (8.18) is obtained. This completes the verification.

Remark. As (8.17) is in vector form, it can be used to obtain the components of stress in a curvilinear coordinate system. For example, in cylindrical polar coordinates R, ϕ, z, the stress on an element whose normal is in the direction of the unit normal vector $\mathbf{e}_R$ is

$$\mathbf{t}(\mathbf{e}_R) = -p\mathbf{e}_R + 2\mu \, \partial \mathbf{v}/\partial R + \mu \mathbf{e}_R \times \text{curl } \mathbf{v}. \tag{8.19}$$

It is conventional to denote the cylindrical polar components of this stress vector by σ_{RR}, $\epsilon_{R\phi}$, σ_{Rz} and hence

$$\mathbf{t}(\mathbf{e}_R) = \sigma_{RR}\mathbf{e}_R + \sigma_{R\phi}\mathbf{e}_\phi + \sigma_{Rz}\mathbf{e}_z. \tag{8.20}$$

Using the methods of §§4.13 and 4.14 [remember that $\mathbf{e}_R$ and $\mathbf{e}_\phi$ are not *constant* vectors], it follows from (8.19) and (8.20) that

$$\sigma_{RR} = -p + 2\mu \, \partial v_R/\partial R$$

$$\sigma_{R\phi} = \mu \left(\frac{\partial v_R}{R\partial\phi} + \frac{\partial v_\phi}{\partial R} - \frac{v_\phi}{R} \right)$$

and

$$\sigma_{Rz} = \mu \left(\frac{\partial v_R}{\partial z} + \frac{\partial v_z}{\partial R} \right).$$

In a similar way, the components of stress on elements whose normals are in the directions of $\mathbf{e}_\phi$ and $\mathbf{e}_z$ may be found.

Isotropic tensors of the fourth order. We shall now prove that any fourth order isotropic tensor U with components u_{ijkm} may be expressed as a sum of products of delta tensors in the form

$$u_{ijkm} = \lambda\delta_{ij}\delta_{km} + \mu\delta_{ik}\delta_{jm} + \nu\delta_{im}\delta_{jk}, \tag{8.21}$$

where λ, μ and ν are arbitrary scalar invariants. The formula will be established by considering the effect of certain rotations of the coordinate axes. We observe first that, for U to be isotropic, under any rotation of the axes

$$u_{pqrs} = l_{pi}l_{qj}l_{rk}l_{sm}u_{ijkm}, \tag{8.22}$$

since the transformed components of U must be the same as those relative to the original axes.

Each of i, j, k and m can assume only the values 1, 2, or 3, and hence the components of U may be collected into the following four groups:

(i)   the components $u_{1111}, u_{2222}, u_{3333}$ in which all four suffixes are equal;

(ii)   components which have three equal suffixes and the other one different (e.g. u_{1112});

(iii)   components which have two distinct pairs of equal suffixes (typically, $u_{1122}, u_{1212}, u_{1221}$);

(iv)   components which have just one pair of equal suffixes and the other suffixes unequal (e.g. u_{1123}).

Consider a rotation of the axes $Ox_1x_2x_3$ through $180°$ about Ox_3. The direction cosines in this transformation are $l_{11} = l_{22} = -1$, $l_{33} = 1$ and $l_{ij} = 0$ when $i \neq j$. Hence, using (8.22),

$$u_{1113} = l_{1i}l_{1j}l_{1k}l_{3m}u_{ijkm} = -u_{1113}$$

and

$$u_{1123} = l_{1i}l_{1j}l_{2k}l_{3m}u_{ijkm} = -u_{1123}.$$

It follows that

$$u_{1113} = 0 \text{ and } u_{1123} = 0.$$

But these are typical components from groups (ii) and (iv), and hence if U is isotropic all components in those groups must vanish.

Consider next a rotation through $90°$ about the axis Ox_3 for which the transformation matrix is:

O	x_1	x_2	x_3
x_1'	0	1	0
x_2'	-1	0	0
x_3'	0	0	1

(8.23)

Using (8.22), under this rotation,

$$u_{1111} = l_{1i}l_{1j}l_{1k}l_{1m}u_{ijkm} = u_{2222}.$$

Similarly, by a cyclic interchange of the suffixes, $u_{2222} = u_{3333}$, and hence the components in group (i) are equal to each other. Also, under the rotation (8.23),

$$u_{1122} = l_{1i}l_{1j}l_{2k}l_{2m}u_{ijkm} = u_{2211}, \tag{8.24}$$

and cyclic interchanges of the suffixes give

$$u_{2233} = u_{3322} \text{ and } u_{3311} = u_{1133}. \qquad (8.25)$$

Now, applying the rotation represented by the transformation matrix

O	x_1	x_2	x_3
x_1'	0	1	0
x_2'	0	0	1
x_3'	1	0	0 ,

we also have

$$u_{1122} = l_{1i}l_{1j}l_{2k}l_{2m}u_{ijkm} = u_{2233}. \qquad (8.26)$$

Advancing the suffixes cyclically and combining the results so obtained with (8.24) and (8.25), it follows that

$$u_{1122} = u_{2211} = u_{2233} = u_{3322} = u_{1133} = u_{3311}.$$

This shows that all members of group (iii) with $i = j$ and $k = m$ are equal to each other. Similarly, it can be shown that all members with $i = k$ and $j = m$ are equal to each other, and that those with $i = m$ and $j = k$ are equal to each other. In summary, the stage now reached is that the tensor with components u_{ijkm} can only be isotropic if:

(a) $u_{1111} = u_{2222} = u_{3333}$;
(b) components such that $i = j \neq k = m$ are equal;
(c) components such that $i = k \neq j = m$ are equal;
(d) components such that $i = m \neq j = k$ are equal;
(e) all components not included under (a)—(d) vanish.

Consider the particular case in which all components in groups (c) and (d) are set equal to zero. A rotation of the axes through 45° about Ox_3 is represented by the transformation matrix:

	x_1	x_2	x_3
x_1'	$\frac{1}{2}\sqrt{2}$	$\frac{1}{2}\sqrt{2}$	0
x_2'	$-\frac{1}{2}\sqrt{2}$	$\frac{1}{2}\sqrt{2}$	0
x_3'	0	0	1

Under this rotation, (8.22) gives

$$u_{1111} = l_{1i}l_{1j}l_{1k}l_{1m}u_{ijkm} = \tfrac{1}{4}(u_{1111} + u_{2222} + u_{1122} + u_{2211}),$$

where the hypothesis that $u_{ijkm} = 0$ except when $i = j$ and $k = m$ has been used. But it has been shown that $u_{1111} = u_{2222}$ and $u_{1122} = u_{2211}$, and it follows therefore that $u_{1111} = u_{1122}$. Hence, all the non-vanishing components of this particular isotropic tensor have the same value, λ, say which may be expressed as

$$u_{ijkm} = \lambda \delta_{ij} \delta_{km}.$$

Similarly, by considering groups (c) and (d), in turn, two other 4th order isotropic tensors may be identified having components

$$u_{ijkm} = \mu \delta_{ik} \delta_{jm}, \quad u_{ijkm} = \nu \delta_{im} \delta_{jk},$$

respectively, where μ and ν are arbitrary scalar invariants. The most general isotropic tensor of order 4 is evidently a combination of all three of these and hence has components of the form

$$u_{ijkm} = \lambda \delta_{ij} \delta_{km} + \mu \delta_{ik} \delta_{jm} + \nu \delta_{im} \delta_{jk},$$

as stated earlier.

EXAMPLE 2. In linear elasticity theory, the material stress components σ_{ij} are assumed to be linear functions of the infinitesimal strain components E_{ij}. Given that σ_{ij} and E_{ij} are the components of second order symmetrical tensors and that the properties of the material are independent of direction, show that

$$\sigma_{ij} = \lambda E_{kk} \delta_{ij} + 2\mu E_{ij}, \tag{8.27}$$

where λ and μ are scalar invariants.

Solution. Denote the coefficient of E_{km} in the linear expression for σ_{ij} by u_{ijkm}, so that

$$\sigma_{ij} = u_{ijkm} E_{km}.$$

By the quotient rule (§8.2 and exercise 4, p. 210), the numbers u_{ijkm} are the components of a fourth-order tensor and the assumption that the properties of the material are independent of direction indicates that this tensor is isotropic. Hence, by (8.21)

$$\sigma_{ij} = \lambda E_{kk} \delta_{ij} + \mu E_{ij} + \nu E_{ji}.$$

But $E_{ji} = E_{ij}$, and hence we may take $\nu = \mu$ giving (8.27).

EXERCISES

5. Using results from the text, prove that:

$$\delta_{ij} \epsilon_{ijk} = 0,$$
$$\epsilon_{ijk} \epsilon_{rjk} = 2\delta_{ir},$$
$$\epsilon_{ijk} \epsilon_{ijk} = 6.$$

6. A second order symmetrical tensor has components s_{ij}. Show that $\epsilon_{ijk} s_{ij} = 0$ for all values of k. Prove the converse, that if S is a second order tensor with components s_{ij} such that $\epsilon_{ijk} s_{ij} = 0$, then S is symmetric.

7. Prove from first principles that $\delta_{ij}\delta_{km}$ are the components of a fourth-order isotropic tensor.

8. Using theorems and results from the text, prove that $\delta_{ij}\epsilon_{kmn}$ are the components of a fifth-order isotropic tensor.

9. Show that the only isotropic tensor of first order is the zero vector. [*Hint.* Consider a 180° rotation about one of the axes.]

10. Prove that the most general isotropic tensor of second order has components $\lambda\delta_{ij}$ where λ is an arbitrary scalar.

11. If A is a second order tensor with components a_{jk}, show from first principles that

$$b_i = \tfrac{1}{2}\epsilon_{ijk}a_{jk} \qquad (i = 1,2,3)$$

are the components of a vector **b**. If the tensor A is anti-symmetrical verify that

$$\mathbf{b} = (a_{23}, a_{31}, a_{12}).$$

12. If A is a second order anti-symmetrical tensor with components a_{jk} and the vector **b** has components given by

$$b_i = \tfrac{1}{2}\epsilon_{ijk}a_{jk} \qquad (i = 1,2,3),$$

show that $a_{rs} = \epsilon_{irs}b_i$. Write down the matrix of components of A in terms of the components of **b**.

13. Prove that

$$\epsilon_{ijk} = \begin{vmatrix} \delta_{i1} & \delta_{i2} & \delta_{i3} \\ \delta_{j1} & \delta_{j2} & \delta_{j3} \\ \delta_{k1} & \delta_{k2} & \delta_{k3} \end{vmatrix}$$

and hence show that

$$\epsilon_{ijk}\epsilon_{rst} = \begin{vmatrix} \delta_{ir} & \delta_{is} & \delta_{it} \\ \delta_{jr} & \delta_{js} & \delta_{jt} \\ \delta_{kr} & \delta_{ks} & \delta_{kt} \end{vmatrix}.$$

Deduce that

$$\epsilon_{ijk}\epsilon_{rsk} = \delta_{ir}\delta_{js} - \delta_{is}\delta_{jr}.$$

14. Using expression (8.14) for the vector product, prove the formula

$$\mathbf{a} \times (\mathbf{b} \times \mathbf{c}) = (\mathbf{a} \cdot \mathbf{c})\mathbf{b} - (\mathbf{a} \cdot \mathbf{b})\mathbf{c}.$$

8.4 Tensor fields

The definitions of scalar and vector fields given in §4.3 extend readily to tensors of general order. If the components $a_{ij\ldots}$ of the tensor A are functions of the coordinates x_1, x_2, x_3 of the points in a given region (or, more generally, any point set), then A is called a *tensor function of position* or a *tensor field.* Tensors which occur in subjects such as fluid dynamics and elasticity theory usually constitute tensor fields and some knowledge of their properties is therefore essential.

The main new feature that we have to consider is the necessity of handling with ease multiple partial derivatives of the components $a_{ij...}$. Such derivatives are often denoted thus:

$$\frac{\partial a_{ij...}}{\partial x_p} = a_{ij...,p}, \quad \frac{\partial^2 a_{ij...}}{\partial x_p \partial x_q} = a_{ij...,pq}, \quad \text{etc.}$$

Particular instances of frequent occurrence are:

$$\frac{\partial a}{\partial x_i} = a_{,i}, \quad \frac{\partial a_i}{\partial x_j} = a_{i,j}, \quad \frac{\partial a_{ij}}{\partial x_k} = a_{ij,k},$$

where i, j and k can usually range over the values 1, 2 or 3.

There is one important basic theorem, which we now state and prove.

Theorem. In an open region $\mathcal{R}$, let the components $a_{ij...}$ (n suffixes) of a tensor A of order n be continuously differentiable functions of the coordinates x_1, x_2, x_3. Then

$$a_{ij...,p} \quad (i, j, ..., p = 1, 2, 3)$$

are the components of a tensor field of order $n+1$ called the *tensor gradient* of A.

Proof. Consider a rotation of the axes $Ox_1 x_2 x_3$ to new positions $Ox_1' x_2' x_3'$, as defined by (8.1). By the tensor law, the transformed components of A are

$$a_{rs...}' = l_{ri} l_{sj} \cdots a_{ij...}.$$

Hence

$$\frac{\partial a_{rs...}'}{\partial x_q'} = l_{ri} l_{sj} \cdots \frac{\partial x_p}{\partial x_q'} \frac{\partial a_{ij...}}{\partial x_p},$$

using the chain rule. But

$$x_p = l_{qp} x_q',$$

and it follows that

$$a_{rs...,q}' = l_{ri} l_{sj} \cdots l_{qp} a_{ij...,p},$$

which is the law of transformation of the components of a tensor of order $n+1$. It is readily verified that the components $a_{ij...,p}$ are unchanged by a translation of the axes (cf. part (a) of the proof that grad Ω is a vector field in §4.4), and hence it follows that they form a tensor field of order $n + 1$ defined on the region $\mathscr{R}$.

The theorem extends to higher order derivatives. The reader should prove, for example, the *corollary* that if the components of A have second order derivatives which are continuous, then $a_{ij...,pq}$ are the components of a tensor of order $n + 2$.

Tensor gradients of scalar fields. One example of a tensor gradient already familiar to us is the vector field grad Ω associated with the scalar Ω. In tensor terminology, Ω is a tensor of order zero and according to the theorem above the components

$$\partial\Omega/\partial x_i = \Omega_{,i} \qquad (i = 1,2,3)$$

form a tensor field of order 1, i.e. a vector field; this is, of course,

$$\text{grad } \Omega = \left(\frac{\partial\Omega}{\partial x_1}, \frac{\partial\Omega}{\partial x_2}, \frac{\partial\Omega}{\partial x_3}\right).$$

By the corollary to the theorem, differentiating Ω twice with respect to the coordinates gives a second order tensor field with components $\Omega_{,ij}$. Contracting this over the two suffixes reduces the order by 2 (see §8.2) and hence yields the scalar field

$$\Omega_{,ii} = \frac{\partial^2\Omega}{\partial x_i \, \partial x_i} = \frac{\partial^2\Omega}{\partial x_1^2} + \frac{\partial^2\Omega}{\partial x_2^2} + \frac{\partial^2\Omega}{\partial x_3^2},$$

which we recognise as the Laplacian $\nabla^2\Omega$.

The tensor gradient of a vector field. The gradient of the vector field with continuously differentiable components F_i is the second order tensor field with components $\partial F_i/\partial x_j$. This tensor may be expressed as a sum of a symmetrical and anti-symmetrical parts by setting

$$\partial F_i/\partial x_j = s_{ij} + a_{ij} \tag{8.28}$$

where

$$s_{ij} = \tfrac{1}{2}\left(\frac{\partial F_i}{\partial x_j} + \frac{\partial F_j}{\partial x_i}\right) \tag{8.29}$$

and

$$a_{ij} = \tfrac{1}{2}\left(\frac{\partial F_i}{\partial x_j} - \frac{\partial F_j}{\partial x_i}\right). \tag{8.30}$$

Two facts worthy of note are the following.

(i) Contracting the symmetrical part s_{ij} over the suffixes i and j gives

$$s_{ii} = \partial F_i/\partial x_i = \text{div } \mathbf{F}.$$

(ii)   Using (8.15),

$$\begin{aligned}(\text{curl } \mathbf{F})_i &= \epsilon_{ijk}\, \partial F_k/\partial x_j \\ &= \epsilon_{ijk} s_{kj} + \epsilon_{ijk} a_{kj}.\end{aligned} \qquad (8.31)$$

But $\epsilon_{ijk} s_{kj} = 0$ (cf. exercise 6, p. 217), and hence taking $i = 1$, 2 and 3 in turn we find that

$$\begin{aligned}\text{curl } \mathbf{F} &= 2(a_{32}, a_{13}, a_{21}) \qquad &(8.32)\\ &= -2(a_{23}, a_{31}, a_{12}). \qquad &(8.33)\end{aligned}$$

This shows that the curl of a vector field is expressible entirely in terms of the non-zero terms of the anti-symmetrical parts of its tensor gradient ($a_{ij} = 0$ when $i = j$, as is seen immediately from the definition (8.30)). Alternatively, we may say that the antisymmetrical part of the tensor gradient of a vector field $\mathbf{F}$ depends only on the components of curl $\mathbf{F}$. More explicitly, setting

$$\text{curl } \mathbf{F} = (c_1, c_2, c_3),$$

and comparing components with those in (8.32) and (8.33), the matrix of components of the anti-symmetrical part of the tensor gradient of $\mathbf{F}$ is

$$\begin{pmatrix} 0 & -\tfrac{1}{2}c_3 & \tfrac{1}{2}c_2 \\ \tfrac{1}{2}c_3 & 0 & -\tfrac{1}{2}c_1 \\ -\tfrac{1}{2}c_2 & \tfrac{1}{2}c_1 & 0 \end{pmatrix}.$$

EXAMPLE 3. In fluid dynamics, the vorticity is defined as $\boldsymbol{\omega} = \text{curl } \mathbf{v}$, where $\mathbf{v}$ is the fluid velocity. If $\mathbf{r} = (x_1, x_2, x_3)$, and a_{ij} are the components of the anti-symmetrical part of the tensor gradient of $\mathbf{v}$, show that the i-th component of the vector field $\tfrac{1}{2}\boldsymbol{\omega} \times \mathbf{r}$ is $a_{ij} x_j$.

Solution. Using (8.14) followed by (8.15), the i-th component of $\tfrac{1}{2}\boldsymbol{\omega} \times \mathbf{r}$ is

$$\begin{aligned}\tfrac{1}{2}(\boldsymbol{\omega} \times \mathbf{r})_i &= \tfrac{1}{2}\epsilon_{ikj}\omega_k x_j \\ &= \tfrac{1}{2}\epsilon_{ikj}\,\epsilon_{krs}(\partial v_s/\partial x_r)x_j.\end{aligned}$$

But an odd number of interchanges of the suffixes of the alternator changes its sign and an even number of interchanges leaves it unchanged. Using this fact followed by (8.13), it follows that

$$\begin{aligned}\tfrac{1}{2}(\boldsymbol{\omega} \times \mathbf{r})_i &= -\tfrac{1}{2}\epsilon_{ijk}\,\epsilon_{rsk}(\partial v_s/\partial x_r)x_j \\ &= \tfrac{1}{2}(\delta_{is}\delta_{jr} - \delta_{ir}\delta_{js})(\partial v_s/\partial x_r)x_j \\ &= \tfrac{1}{2}\left(\frac{\partial v_i}{\partial x_j} - \frac{\partial v_j}{\partial x_i}\right)x_j \\ &= a_{ij} x_j.\end{aligned}$$

This is the required result.

EXERCISES

15. Use tensor methods to establish the following identities:

(i) $\operatorname{div}(\mathbf{F} \times \mathbf{G}) \equiv \mathbf{G} . \operatorname{curl} \mathbf{F} - \mathbf{F} . \operatorname{curl} \mathbf{G}$;

(ii) $\operatorname{curl}(\Omega \mathbf{F}) \equiv \Omega \operatorname{curl} \mathbf{F} - \mathbf{F} \times \operatorname{grad} \Omega$;

(iii) $\operatorname{curl} \operatorname{curl} \mathbf{F} \equiv \operatorname{grad} \operatorname{div} \mathbf{F} - \nabla^2 \mathbf{F}$.

In each case, assume that all derivatives which are implied by these relations exist and are continuous.

8.5 The divergence theorem in tensor field theory

The gradient of a second order tensor field A with continuously differentiable components a_{ij} is a third order tensor with components $a_{ij,k}$. Contraction over the suffixes j and k gives a vector field whose i-th component is

$$a_{ij,j} = \frac{\partial a_{i1}}{\partial x_1} + \frac{\partial a_{i2}}{\partial x_2} + \frac{\partial a_{i3}}{\partial x_3}, \qquad (8.34)$$

whilst interchanging i and j (for convenience) and then contracting over the suffixes j and k gives another vector field whose i-th component is

$$a_{ji,j} = \frac{\partial a_{1i}}{\partial x_1} + \frac{\partial a_{2i}}{\partial x_2} + \frac{\partial a_{3i}}{\partial x_3}. \qquad (8.35)$$

In the special case when A is symmetrical, the two vector fields will, of course, be identical since then $a_{ij} = a_{ji}$. If $\mathbf{F}$ is a vector field with continuous first derivatives, div $\mathbf{F}$ may be obtained by contracting the components $F_{i,j}$ of the tensor gradient of $\mathbf{F}$ with respect to the suffixes i, j, and hence (8.34) and (8.35) may be regarded as generalised divergences; they are, in fact, sometimes called the divergence of A with respect to the second suffix and the divergence of A with respect to the first suffix, respectively.

The divergence theorem proved in Chapter 6 may be extended to obtain corresponding results for the vector fields (8.34) and (8.35). Using our present notation, the divergence theorem (6.1) may be expressed as

$$\int_S F_i n_i \, dS = \int_\tau F_{i,i} \, d\tau, \qquad (8.36)$$

where $\mathbf{F} = (F_1, F_2, F_3)$ is a continuously differentiable vector field defined in a region τ bounded by a simple closed surface S and $\mathbf{n} = (n_1, n_2, n_3)$ is the outward unit normal vector to S. To generalize the theorem, consider the three results (6.3), (6.4), and (6.5), from which (8.36) is derived. Replacing F_1, F_2, F_3, respectively by the continuously differentiable scalar functions a_{i1}, a_{i2}, a_{i3} and making an appropriate change of notation for the coordinates and unit vectors, we have

$$\int_S a_{i1}\,\mathbf{e}_1 \cdot \mathbf{dS} = \int_\tau a_{i1,1}\,d\tau$$

$$\int_S a_{i2}\,\mathbf{e}_2 \cdot \mathbf{dS} = \int_\tau a_{i2,2}\,d\tau$$

$$\int_S a_{i3}\,\mathbf{e}_3 \cdot \mathbf{dS} = \int_\tau a_{i3,3}\,d\tau.$$

Addition of these results gives

$$\int_S a_{ij}n_j\,dS = \int_\tau a_{ij,j}\,d\tau, \tag{8.37}$$

where $n_j = \mathbf{e}_j \cdot \mathbf{n}$. This is one of the generalised forms of the divergence theorem (8.36). Another form, corresponding to (8.35), is

$$\int_S a_{ji}n_j\,dS = \int_\tau a_{ji,j}\,d\tau. \tag{8.38}$$

Further generalisation is possible for tensors of higher order, but the cases we have mentioned are sufficient for many purposes.

EXERCISES

16. In the region τ bounded by a simple closed surface S, the second order *symmetrical* tensor field with components σ_{ij} has continuous first order derivatives and is such that, throughout τ, $\sigma_{ij,j} = 0$. On the bounding surface S, a vector field $\mathbf{F}$ is defined with i-th component $\sigma_{ij}n_j$, where $\mathbf{n} = (n_1, n_2, n_3)$ is the outward unit normal vector to S. Prove that:

$$\text{(i)} \quad \int_S \mathbf{F}\,dS = 0, \qquad\qquad \text{(ii)} \quad \int_S \mathbf{r} \times \mathbf{F}\,dS = 0,$$

where $\mathbf{r} = (x_1, x_2, x_3)$ is the position vector from the origin.

17. In the region τ bounded by a simple closed surface S, the vector field $\mathbf{v} = (v_1, v_2, v_3)$ is continuously differentiable and the components of the symmetrical part of its tensor gradient are denoted by e_{ij}. If the tensor with components σ_{ij} is symmetrical and has continuous first order derivatives in τ such that $\sigma_{ij,j} = 0$, show that

$$\int_S v_i\sigma_{ij}n_j\,dS = \int_\tau \sigma_{ij}e_{ij}\,d\tau.$$

CHAPTER 9

REPRESENTATION THEOREMS FOR
ISOTROPIC TENSOR FUNCTIONS

9.1 Introduction

In theoretical studies of the behaviour of materials it is necessary to introduce postulates about the way in which materials respond to factors such as local deformations and local temperature gradients. These postulates naturally depend on the type of material under consideration and in continuum theories they are expressed mathematically in the form of certain relations between tensors called *constitutive equations*. A simple example of a constitutive equation is that relating the stress components σ_{ij} and the pressure p in an inviscid fluid, viz.

$$\sigma_{ij} = -p\delta_{ij}.$$

Another example is the constitutive equation of linear elasticity theory, which is expressed in the form of the relation (8.27); we observe that it is a linear relation between the components σ_{ij} of the stress tensor and the infinitesimal strain components E_{ij}. Yet another example of a constitutive equation is Fourier's law of heat conduction, which may be expressed as

$$\mathbf{q} = -k \operatorname{grad} T,$$

where $\mathbf{q}$ is the heat conduction vector, T denotes temperature and k is the thermal conductivity.

Now a basic postulate which is applicable to a large class of materials is that the physical properties are independent of direction, or, in other words, that there are no preferred directions of response. Such materials are said to be *isotropic*. Mathematically, the requirement of isotropy implies that the relations defining the material properties should be invariant under a rotation of the coordinate axes, and this places certain restrictions on the form that the relations may take.

The main purpose of this chapter is to consider three types of relation which are of particular importance, and to derive theorems for their representation under the assumption of isotropy. Many generalisations of these theorems have been proved in recent years but their applications are largely at postgraduate level and lie outside the scope of the present book. Two of the cases considered involve second order symmetrical tensors and it is necessary to establish first some of the principal properties of such tensors.

9.2 Diagonalization of second-order symmetrical tensors

A classical theorem of great importance is that for any second order symmetrical cartesian tensor, it is always possible to choose axes, called *principal axes*, such that the matrix of components of the tensor is of the *diagonal form*

$$\begin{pmatrix} \lambda_1 & 0 & 0 \\ 0 & \lambda_2 & 0 \\ 0 & 0 & \lambda_3 \end{pmatrix}.$$

This section is devoted to proving this theorem.

DEFINITION. A non-zero vector $\mathbf{v} = (v_1, v_2, v_3)$ is said to be an *eigenvector* of the second order tensor A, with components a_{ij}, if there exists a scalar λ such that

$$a_{ij} v_j = \lambda v_i \qquad (i = 1, 2, 3); \tag{9.1}$$

λ is called the *eigenvalue* associated with the vector $\mathbf{v}$.

Written out in full, equations (9.1) are:

$$(a_{11} - \lambda)v_1 + a_{12} v_2 + a_{13} v_3 = 0$$
$$a_{21} v_1 + (a_{22} - \lambda)v_2 + a_{23} v_3 = 0$$
$$a_{31} v_1 + a_{32} v_2 + (a_{33} - \lambda)v_3 = 0.$$

A necessary and sufficient condition for these equations to have a solution in which at least one of v_1, v_2, v_3 is non-zero is that

$$\begin{vmatrix} a_{11} - \lambda & a_{12} & a_{13} \\ a_{21} & a_{22} - \lambda & a_{23} \\ a_{31} & a_{32} & a_{33} - \lambda \end{vmatrix} = 0. \tag{9.2}$$

This is a cubic equation for λ, called the *characteristic equation* of A. It follows that there are three eigenvalues $\lambda_1, \lambda_2, \lambda_3$ (though these may not be distinct), and corresponding to these eigenvalues there will be eigenvectors, which may be denoted by $\mathbf{v}^{(1)}, \mathbf{v}^{(2)}, \mathbf{v}^{(3)}$, respectively. It should be noted that the eigenvectors are arbitrary to the extent of a multiplicative scalar invariant factor. In other words, if $\mathbf{v}$ is an eigenvector and $\gamma \neq 0$ is a scalar invariant, then $\gamma \mathbf{v}$ is also an eigenvector.

THEOREM 1. The eigenvalues of a second order tensor are independent of the coordinate system.

Proof. With respect to the axes $Ox_1'x_2'x_3'$, suppose that λ is an eigenvalue of the tensor A, and hence that

$$a_{rs}'v_s' = \lambda v_r'.$$

Expressing a_{rs}', v_s' and v_r' in terms of components relative to the axes $Ox_1x_2x_3$,

$$l_{ri}l_{sj}a_{ij}l_{sp}v_p = \lambda l_{rq}v_q.$$

Multiplying by l_{rt} and using the orthonormality conditions, we obtain

$$\delta_{it}a_{ij}\delta_{jp}v_p = \lambda\delta_{tq}v_q,$$

which reduces to

$$a_{tj}v_j = \lambda v_t.$$

Thus λ is also an eigenvalue in the coordinate system $Ox_1x_2x_3$, and the theorem follows.

THEOREM 2. A real symmetric second order tensor has real eigenvalues.

Proof. Let λ be any eigenvalue of the tensor A. Multiplying (9.1) by the complex conjugate of v_i (which is denoted by $\bar{v}_i$),

$$\begin{aligned}
\lambda v_i \bar{v}_i &= a_{ij}v_j\bar{v}_i \\
&= \tfrac{1}{2}a_{ij}v_j\bar{v}_i + \tfrac{1}{2}a_{ji}v_j\bar{v}_i \quad \text{(because } A \text{ is symmetrical)} \\
&= \tfrac{1}{2}a_{ij}(v_j\bar{v}_i + v_i\bar{v}_j).
\end{aligned}$$

The expression $v_j\bar{v}_i + \bar{v}_j v_i$ is real because it is the sum of conjugates, and since $v_i\bar{v}_i$ is real and non-zero, we conclude that λ is real.

THEOREM 3. If A is a real symmetric second order tensor, it is possible to choose a set of principal axes, say $Ox_1''x_2''x_3''$, relative to which the components of A are

$$a_{11}'' = \lambda_1, \quad a_{22}'' = \lambda_2, \quad a_{33}'' = \lambda_3 \quad \text{and} \quad a_{ij}'' = 0 \quad \text{for } i \neq j,$$

where $\lambda_1, \lambda_2, \lambda_3$ are the eigenvalues of A.

Proof. Let λ_1 be an eigenvalue of A, and denote by $\mathbf{v}^{(1)}$ a corresponding eigenvector. Then

$$a_{ij}v_j^{(1)} = \lambda_1 v_i^{(1)}.$$

If $\mathbf{e}^{(1)}$ is the unit vector parallel to $\mathbf{v}^{(1)}$, it follows also that

$$a_{ij}e_j^{(1)} = \lambda_1 e_i^{(1)},$$

because $\mathbf{e}^{(1)} = \mathbf{v}^{(1)}/|\mathbf{v}^{(1)}|$.

Choose axes $Ox_1'x_2'x_3'$ such that Ox_1' is parallel to $\mathbf{e}^{(1)}$. Then, in the transformation matrix

	x_1	x_2	x_3
x_1'	l_{11}	l_{12}	l_{13}
x_2'	l_{21}	l_{22}	l_{23}
x_3'	l_{31}	l_{32}	l_{33}

,

$l_{11} = e_1^{(1)}$, $l_{12} = e_2^{(1)}$ and $l_{13} = e_3^{(1)}$.

Now

$$a'_{rs} = l_{ri} l_{sj} a_{ij},$$

and hence

$$\begin{aligned}
a'_{r1} &= l_{ri} l_{1j} a_{ij} \\
&= l_{ri} e_j^{(1)} a_{ij} \\
&= \lambda_1 l_{ri} e_i^{(1)} \\
&= \lambda_1 l_{ri} l_{1i} \\
&= \lambda_1 \delta_{r1}.
\end{aligned}$$

Since A is symmetrical in all coordinate systems (see §8.2), it follows that, relative to the axes $Ox'_1 x'_2 x'_3$, the matrix of components of A is of the form

$$\begin{pmatrix} \lambda_1 & 0 & 0 \\ 0 & a'_{22} & a'_{23} \\ 0 & a'_{32} & a'_{33} \end{pmatrix}$$

Suppose now that the axes $Ox'_1 x'_2 x'_3$ are rotated about Ox'_1, and denote the axes in their new position by $Ox''_1 x''_2 x''_3$. Since Ox'_1 and Ox''_1 coincide, the transformation matrix is of the form

	x'_1	x'_2	x'_3
x''_1	1	0	0
x''_2	0	l'_{22}	l'_{23}
x''_3	0	l'_{32}	l'_{33}

By considering the equations

$$a''_{rs} = l'_{ri} l'_{sj} a_{ij},$$

it is readily verified that this rotation of the axes changes only those components of A having suffixes 22, 23, 32, and 33, respectively. The completion of the proof therefore rests upon showing that it is possible to rotate the axes about Ox'_1 in such a way that the set of scalars

$$\begin{pmatrix} a'_{22} & a'_{23} \\ a'_{32} & a'_{33} \end{pmatrix}$$

transforms to a set

$$\begin{pmatrix} \lambda_2 & 0 \\ 0 & \lambda_3 \end{pmatrix}.$$

Now this is the two-dimensional analogue of the theorem we require to prove, and by reasoning similar to that above (with all suffixes restricted to have two values instead of three), its proof depends in turn, upon proving the one-dimensional analogue. But the one-dimensional analogue is trivially true, because the matrix representation of the tensor is then a single scalar whose eigenvalue is itself. We conclude therefore that it *is* possible to choose axes $Ox_1''x_2''x_3''$ relative to which the tensor A has components given by the matrix

$$\begin{pmatrix} \lambda_1 & 0 & 0 \\ 0 & \lambda_2 & 0 \\ 0 & 0 & \lambda_3 \end{pmatrix}.$$

The diagonal elements are the eigenvalues of A, which, by virtue of Theorem 1, are independent of the choice of axes. This completes the proof.

THEOREM 4. Let $Ox_1'x_2'x_3'$ be a set of rectangular cartesian coordinate axes such that Ox_1', Ox_2', Ox_3' are in the directions of the unit vectors $\mathbf{e}^{(1)}$, $\mathbf{e}^{(2)}$, $\mathbf{e}^{(3)}$, respectively. If $\mathbf{e}^{(1)}$, $\mathbf{e}^{(2)}$, $\mathbf{e}^{(3)}$ are eigenvectors of the second order tensor A, corresponding to the eigenvalues λ_1, λ_2, λ_3, respectively, then, relative to the axes $Ox_1'x_2'x_3'$, the components of A are $a_{11}' = \lambda_1$, $a_{22}' = \lambda_2$, $a_{33}' = \lambda_3$ and $a_{rs}' = 0$ if $r \neq s$.

Proof. Let A have components a_{ij} relative to the axes $Ox_1x_2x_3$. Then by hypothesis,

$$a_{ij}e_j^{(s)} = \lambda_s e_i^{(s)}, \qquad (s = 1, 2, 3)$$

where the expression on the right-hand side is *not* summed over s.

The transformation matrix relating the coordinate systems $Ox_1\dot{x}_2x_3$ and $Ox_1'x_2'x_3'$ has components $l_{ij} = e_j^{(i)}$. Hence, relative to the axes $Ox_1'x_2'x_3'$, A has components

$$\begin{aligned} a_{rs}' = l_{ri}l_{sj}a_{ij} &= e_i^{(r)}e_j^{(s)}a_{ij} \\ &= e_i^{(r)}\lambda_s e_i^{(s)} \\ &= \delta_{rs}\lambda_s \quad \text{(not summed over } s\text{)}, \end{aligned}$$

which yields the required results.

EXAMPLE 1. Find three orthogonal eigenvectors of the tensor whose matrix of components is

$$\begin{pmatrix} 0 & 0 & 0 \\ 0 & 1 & 1 \\ 0 & 1 & 1 \end{pmatrix}.$$

Solution. The eigenvectors are obtained by finding the non-trivial solutions of the set of linear equations

$$\begin{aligned} -\lambda v_1 &= 0 \\ (1 - \lambda)v_2 + v_3 &= 0 \\ v_2 + (1 - \lambda)v_3 &= 0. \end{aligned} \tag{9.3}$$

The characteristic equation is

$$\begin{vmatrix} -\lambda & 0 & 0 \\ 0 & 1-\lambda & 1 \\ 0 & 1 & 1-\lambda \end{vmatrix} = 0,$$

which reduces to

$$\lambda[(1-\lambda)^2 - 1] = 0.$$

Solving, we obtain the three eigenvalues

$$\lambda = 2, \quad \lambda = 0, \quad \lambda = 0.$$

Taking $\lambda = 2$, equations (9.3) reduce to

$$v_1 = 0 \quad \text{and} \quad v_2 = v_3.$$

Hence one eigenvector is

$$\mathbf{v}^{(1)} = (0, 1, 1).$$

Taking $\lambda = 0$, the first of equations (9.3) is satisfied for all values of v_1, and the other two equations reduce to $v_2 = -v_3$. Hence all vectors of the form $(v_1, v_2, -v_2)$ are eigenvectors orthogonal to $\mathbf{v}^{(1)}$. Thus we may choose, say,

$$\mathbf{v}^{(2)} = (0, -1, 1)$$

and

$$\mathbf{v}^{(3)} = (1, 0, 0).$$

These vectors are orthogonal to each other and orthogonal to $\mathbf{v}^{(1)}$, as required.

EXERCISE

1. Find the eigenvalues and eigenvectors of the tensor A whose components a_{ij} relative to rectangular cartesian axes $Ox_1x_2x_3$ are

$$\begin{pmatrix} 0 & 0 & 0 \\ 0 & 1 & 2 \\ 0 & 2 & 1 \end{pmatrix}.$$

Find an orthonormal set S of unit vectors such that, when referred to S, A becomes

$$\begin{pmatrix} 0 & 0 & 0 \\ 0 & -1 & 0 \\ 0 & 0 & 3 \end{pmatrix}.$$

9.3   Invariants of second order symmetrical tensors

By Theorem 1 of the previous section, the eigenvalues λ_1, λ_2, λ_3 of a second order symmetrical tensor A are independent of the coordinate system and they are therefore said to be *invariants* of A. However, it should be noted that the eigenvalues are not an ordered set, although they can be made so if they are distinct by requiring, for example, that the largest appear first and the smallest last.

Any function $f(\lambda_1, \lambda_2, \lambda_3)$ whose value is unaltered by interchanges of pairs of λ_1, λ_2, λ_3 is said to be *symmetric* in λ_1, λ_2, λ_3 and is an invariant of the tensor A. Such invariants play an important role in the development of theoretical continuum mechanics. Particular examples are the elementary symmetric functions

$$I_1 = \lambda_1 + \lambda_2 + \lambda_3, \tag{9.4}$$
$$I_2 = \lambda_1\lambda_2 + \lambda_2\lambda_3 + \lambda_3\lambda_1, \tag{9.5}$$

and

$$I_3 = \lambda_1\lambda_2\lambda_3; \tag{9.6}$$

they are called the *principal invariants* of A.

The characteristic equation (9.2) which determines the eigenvalues of A can be expressed as

$$(\lambda - \lambda_1)(\lambda - \lambda_2)(\lambda - \lambda_3) = 0$$

because its roots are λ_1, λ_2, λ_3. Expanding the left-hand side and using the definitions (9.4)–(9.6), we find that

$$\lambda^3 - I_1\lambda^2 + I_2\lambda - I_3 = 0, \tag{9.7}$$

which shows that the coefficients of the characteristic equation can be expressed in terms of the principal invariants of A.

Symmetric functions of the eigenvalues. Any symmetric function $f(\lambda_1, \lambda_2, \lambda_3)$ of the eigenvalues of A can be expressed as a function of the coefficients I_1, I_2, I_3 in the characteristic equation, for Cardan's solution of the cubic equation (9.7) identifies the roots λ_1, λ_2, λ_3 as functions of I_1, I_2, and I_3. By substituting Cardan's expressions into $f(\lambda_1, \lambda_2, \lambda_3)$, and observing that the value obtained will be independent of the ordering of the roots because of the assumed symmetry, we obtain, as required

$$f(\lambda_1, \lambda_2, \lambda_3) = g(I_1, I_2, I_3).$$

EXERCISES

2. If λ_1 λ_2 and λ_3 are the roots of equation (9.7), express

(i) $I = \lambda_1^2\lambda_2\lambda_3 + \lambda_1\lambda_2^2\lambda_3 + \lambda_1\lambda_2\lambda_3^2$

and

(ii) $\mathcal{J} = \lambda_1^3\lambda_2\lambda_3 + \lambda_1\lambda_2^3\lambda_3 + \lambda_1\lambda_2\lambda_3^3$

as functions of I_1, I_2, and I_3.

3. A second order symmetrical tensor A has components a_{ij}. Using only the tensor transformation law and the orthonormality relations, show that the scalars

$$\mathcal{J}_1 = a_{ii}, \quad \mathcal{J}_2 = a_{ij}a_{ji}, \quad \mathcal{J}_3 = a_{ij}a_{jk}a_{ki}$$

are invariants of A. By referring A to its principal axes express $\mathcal{J}_1$, $\mathcal{J}_2$ and $\mathcal{J}_3$ in terms of the eigenvalues λ_1, λ_2 and λ_3 of A.

The principal invariants of A are I_1, I_2 and I_3. Verify that

$$I_1 = \mathcal{J}_1, \quad I_2 = \tfrac{1}{2}(\mathcal{J}_1^2 - \mathcal{J}_2), \quad I_3 = \tfrac{1}{6}(2\mathcal{J}_3 + \mathcal{J}_1^3 - 3\mathcal{J}_1\mathcal{J}_2).$$

9.4 Representation of isotropic vector functions

Relative to the rectangular cartesian coordinate axes $Ox_1x_2x_3$, let the vectors $\mathbf{a}$ and $\mathbf{b}$ have components (a_1, a_2, a_3) and (b_1, b_2, b_3), respectively, and suppose that the components of $\mathbf{b}$ are functions of the components of $\mathbf{a}$. Then

$$b_i = F_i(a_1, a_2, a_3) \qquad (i = 1, 2, 3). \tag{9.8}$$

Suppose that, relative to the axes $Ox_1'x_2'x_3'$ whose orientations relative to the original axes $Ox_1x_2x_3$ are defined by the usual transformation matrix (8.1), the components of $\mathbf{a}$ and $\mathbf{b}$ are (a_1', a_2', a_3') and (b_1', b_2', b_3'), respectively, and that

$$b_i' = F_i(a_1', a_2', a_3'), \qquad (i = 1, 2, 3), \tag{9.9}$$

where the *functions* F_i are *the same as before*. Then the dependence of the components of $\mathbf{b}$ on the components of $\mathbf{a}$ is unchanged by a rotation of the axes and $\mathbf{b}$ is said to be an *isotropic function* of $\mathbf{a}$.

The condition of isotropy restricts severely the form that the functions F_i can take, as the following theorem shows.

Theorem. If the vector $\mathbf{b}$ is an isotropic function of the vector $\mathbf{a}$, then it may be represented in the form

$$\mathbf{b} = \lambda(a)\mathbf{a}, \tag{9.10}$$

where $\lambda(a)$ is a scalar invariant function of $a = |\mathbf{a}|$. Conversely, if (9.10) holds, then $\mathbf{b}$ is an isotropic function of $\mathbf{a}$.

Proof. Suppose that $\mathbf{b}$ is an isotropic function of $\mathbf{a}$. Choose axes $Ox_1x_2x_3$ such that $\mathbf{a}$ is parallel to Ox_1 and hence has components $(a_1, 0, 0)$, and let the corresponding components of $\mathbf{b}$ be (b_1, b_2, b_3). Then

$$b_1 = F_1(a_1, 0, 0), \quad b_2 = F_2(a_1, 0, 0), \quad b_3 = F_3(a_1, 0, 0). \tag{9.11}$$

Let the axes $Ox_1x_2x_3$ be rotated through $180°$ about Ox_1. The new axes $Ox_1'x_2'x_3'$ will then be such that Ox_1' is along Ox_1, and Ox_2' and Ox_3' are in the opposite directions to Ox_2 and Ox_3, respectively. Relative to the new axes, the components of $\mathbf{a}$ and $\mathbf{b}$ will therefore be $(a_1, 0, 0)$ and $(b_1, -b_2, -b_3)$, respectively, and since $\mathbf{b}$ is assumed to be an isotropic function of $\mathbf{a}$ we must have

$$b_1 = F_1(a_1, 0, 0), \quad -b_2 = F_2(a_1, 0, 0), \quad -b_3 = F_3(a_1, 0, 0). \tag{9.12}$$

Comparing (9.11) and (9.12), it is seen that $b_2 = b_3 = 0$.

Consider next a rotation of the axes $Ox_1x_2x_3$ through $180°$ about Ox_2 to new positions $Ox_1'x_2'x_3'$. The axis Ox_1' will be opposite to Ox_1, and hence relative to the new axes, $\mathbf{a} = (-a_1, 0, 0)$ and $\mathbf{b} = (-b_1, 0, 0)$. Because of the assumption of isotropy,

$$-b_1 = F_1(-a_1, 0, 0)$$

and using (9.11) it follows that

$$F_1(-a_1, 0, 0) = -F_1(a_1, 0, 0).$$

Changing the sign of a_1 is therefore equivalent to changing the sign of F_1, and hence $F_1(a_1)$ must be expressible in the form $a_1 \lambda(|a_1|)$, where λ is a function of $|a_1|$ only. But as $a_2 = a_3 = 0$, $|a_1| = a$, and hence we conclude that

$$b_1 = \lambda(a)a_1, \quad b_2 = a_2 = 0, \quad b_3 = a_3 = 0.$$

Thus

$$\mathbf{b} = \lambda(a)\mathbf{a},$$

as required.

To prove that whenever the relation (9.10) holds $\mathbf{b}$ is an isotropic function of $\mathbf{a}$, consider the components

$$b_i = \lambda(a)a_i \qquad (i = 1, 2, 3). \tag{9.13}$$

Multiplying each side of this equation by l_{ji} (defined by (8.1)) and summing over the three values of i gives

$$l_{ji}b_i = \lambda(a)l_{ji}a_i.$$

By the rule for transforming the components of a vector, this may be expressed as

$$b'_j = \lambda(a)a'_j \qquad (j = 1, 2, 3) \tag{9.14}$$

which is the j-th component of (9.10) relative to the axes $Ox'_1x'_2x'_3$. The functional dependence of the components of $\mathbf{b}$ on the components of $\mathbf{a}$ is therefore unaffected by a rotation of the axes, and hence $\mathbf{b}$ is an isotropic function of $\mathbf{a}$.

EXERCISE

4. If n is a constant, determine for which values of n the following vector function $\mathbf{c}$ of (a_1, a_2, a_3) is isotropic:

$$\mathbf{c} = (c_1, c_2, c_3)$$

where

$$c_1 = a_1^{n+1} + a_2^n a_1 + a_3^n a_1,$$
$$c_2 = a_1^n a_2 + a_2^{n+1} + a_2^n a_2$$

and

$$c_3 = a_1^n a_3 + a_2^n a_3 + a_3^{n+1}.$$

9.5 Isotropic scalar functions of symmetrical second order tensors

Relative to the rectangular cartesian coordinate axes $Ox_1x_2x_3$, let the second order symmetrical tensor A have components a_{ij} $(i,j = 1, 2, 3)$, and suppose that Ω is a scalar function of these components. For brevity, the relation may be expressed as

$$\Omega = \Omega(a_{ij}),$$

where it is to be understood that Ω is a function of *all* the components of A. Relative to new axes $Ox'_1x'_2x'_3$, suppose that the components of A are a'_{ij} and that

$$\Omega(a'_{ij}) = \Omega(a_{ij}). \tag{9.15}$$

This indicates that the dependence of Ω on the components of A is unchanged by a rotation of the axes, and Ω is therefore said to be an *isotropic function* of A. The theorem which follows shows that such isotropic scalar functions can be represented in a relatively simple form.

Theorem. If Ω is an isotropic scalar function of the second order symmetrical tensor A, then it may be represented in the form

$$\Omega = \Omega(I_1, I_2, I_3), \tag{9.16}$$

where I_1, I_2, I_3 are the principal invariants of A (defined by relations (9.4)–(9.6)). Conversely, if (9.16) holds, then Ω is an isotropic function of A.

Proof. Suppose that Ω is an isotropic function of A. As A is a second order symmetrical tensor, Theorem 3 of §9.2 shows that a principal set of axes $Ox_1x_2x_3$ may be chosen relative to which A has the diagonal form

$$\begin{pmatrix} \lambda_1 & 0 & 0 \\ 0 & \lambda_2 & 0 \\ 0 & 0 & \lambda_3 \end{pmatrix}. \tag{9.17}$$

In general Ω depends on $a_{11}, a_{22}, a_{33}, a_{12}, a_{23}$ and a_{31} (because of the symmetry of A it is unnecessary to list a_{21}, a_{32} and a_{13}), and hence relative to the principal axes $Ox_1x_2x_3$,

$$\Omega = \Omega(\lambda_1, \lambda_2, \lambda_3, 0, 0, 0).$$

Consider a rotation to new axes $Ox_1'x_2'x_3'$, defined by the transformation matrix:

O	x_1	x_2	x_3
x_1'	0	1	0
x_2'	-1	0	0
x_3'	0	0	1

$$(9.18)$$

This rotation turns the axes through $90°$ about Ox_3. Using the tensor transformation rule, viz.

$$a_{rs}' = l_{ri}l_{sj}a_{ij},$$

and noting that the components a_{ij} are as displayed in the matrix (9.17),

$$a_{11}' = l_{1i}l_{1j}a_{ij} = \lambda_2,$$
$$a_{22}' = l_{2i}l_{2j}a_{ij} = \lambda_1,$$

and

$$a_{33}' = l_{3i}l_{3j}a_{ij} = \lambda_3.$$

It is readily verified that $a_{ij}' = 0$ when $i \neq j$, and hence relative to the axes $Ox_1'x_2'x_3'$ A has the diagonal form

$$\begin{pmatrix} \lambda_2 & 0 & 0 \\ 0 & \lambda_1 & 0 \\ 0 & 0 & \lambda_3 \end{pmatrix}.$$

By the assumed condition of isotropy, it follows that

$$\Omega(\lambda_2, \lambda_1, \lambda_3, 0, 0, 0) = \Omega(\lambda_1, \lambda_2, \lambda_3, 0, 0, 0)$$

and hence interchanging λ_1 and λ_2 leaves the value of Ω unchanged. By a similar method it may be shown that interchanging any pair of the eigenvalues λ_1, λ_2, λ_3 leaves the value of Ω unchanged, and hence Ω is a symmetric function of λ_1, λ_2 and λ_3. From the remarks at the end of §9.3, p. 230, we conclude that Ω is a function of the principal invariants I_1, I_2, I_3 of A, as required.

To establish that when (9.16) holds Ω is an isotropic function of A is straightforward. For, I_1, I_2 and I_3 are invariant under a rotation of the axes and hence if (9.16) holds for one set of axes it holds for all sets obtained by a rotation. The dependence of Ω on the components of A is therefore invariant under a rotation of the axes and hence Ω is an isotropic function of A.

9.6 Representation of an isotropic tensor function

Let A and B be second order symmetrical tensors whose components relative to the axes $Ox_1x_2x_3$ are a_{km} and b_{ij}, respectively. Suppose that each of the components of B is a function of the components of A, and denote this by writing

$$b_{ij} = F_{ij}(a_{km}). \qquad (9.19)$$

Relative to new axes $Ox'_1x'_2x'_3$, suppose that the components of A and B are a'_{km} and b'_{ij}, respectively, and that the relations

$$b'_{ij} = F_{ij}(a'_{km})$$

are satisfied, where the functions F_{ij} are as in (9.19). The dependence of the components of B on the components of A is then invariant under a rotation of the axes and B is said to be an *isotropic function* of A. As in the case of isotropic vector functions and isotropic scalar functions of a tensor, discussed in the previous two sections of this chapter, the condition of isotropy is naturally satisfied by only a limited class of functions. The final representation theorem that we prove shows the extent of the simplification that can be achieved.

Theorem. Let A be a second order symmetrical tensor whose principal invariants are I_1, I_2 and I_3. If the second order symmetrical tensor B is an isotropic function of A, then there exist scalar invariant functions α, β and γ of I_1, I_2 and I_3 such that the components b_{ij} of B can be represented in terms of the components a_{ij} of A in the form

$$b_{ij} = \alpha\delta_{ij} + \beta a_{ij} + \gamma a_{ik}a_{kj} \qquad (i,j = 1,2,3). \qquad (9.20)$$

Conversely, if (9.20) holds, then B is an isotropic function of A.

Proof. Suppose that B is an isotropic function of A and choose the axes $Ox_1x_2x_3$ such that A is of the diagonal form

$$\begin{pmatrix} \lambda_1 & 0 & 0 \\ 0 & \lambda_2 & 0 \\ 0 & 0 & \lambda_3 \end{pmatrix}. \qquad (9.21)$$

Then

$$b_{ij} = F_{ij}(\lambda_1, \lambda_2, \lambda_3, 0, 0, 0). \qquad (9.22)$$

Consider a rotation of the axes through $180°$ about Ox_1, defined by the transformation matrix

O	x_1	x_2	x_3
x_1'	1	0	0
x_2'	0	-1	0
x_3'	0	0	-1.

Let the components of A and B relative to the new axes $Ox_1'x_2'x_3'$ be a_{ij}' and b_{ij}', respectively. Using the law for transforming the components of a second order tensor,

$$a_{11}' = l_{1i}l_{1j}a_{ij} = \lambda_1,$$
$$a_{22}' = l_{2i}l_{2j}a_{ij} = \lambda_2,$$
$$a_{33}' = l_{3i}l_{3j}a_{ij} = \lambda_3,$$

and

$$a_{rs}' = l_{ri}l_{sj}a_{ij} = 0 \quad \text{when } r \neq s.$$

Also,

$$b_{12}' = l_{1i}l_{2j}b_{ij} = -b_{12}$$

and

$$b_{13}' = l_{1i}l_{3j}b_{ij} = -b_{13}.$$

Now from (9.22),

$$b_{12} = F_{12}(\lambda_1, \lambda_2, \lambda_3, 0, 0, 0) \tag{9.23}$$

and

$$b_{13} = F_{13}(\lambda_1, \lambda_2, \lambda_3, 0, 0, 0). \tag{9.24}$$

Under the assumed condition of isotropy, the functions F_{12} and F_{13} are unchanged by a rotation of the axes, and hence relative to the axes $Ox_1'x_2'x_3'$,

$$-b_{12} = F_{12}(\lambda_1, \lambda_2, \lambda_3, 0, 0, 0) \tag{9.25}$$
$$-b_{13} = F_{13}(\lambda_1, \lambda_2, \lambda_3, 0, 0, 0). \tag{9.26}$$

Comparing these relations with (9.23) and (9.24), it is seen that $b_{12} = b_{13} = 0$. In a similar way, it may be shown that $b_{23} = 0$, and hence relative to the axes $Ox_1x_2x_3$, B is of the diagonal form

$$\begin{pmatrix} \mu_1 & 0 & 0 \\ 0 & \mu_2 & 0 \\ 0 & 0 & \mu_3 \end{pmatrix},$$

where μ_1, μ_2, μ_3 are the eigenvalues of B. One implication of the assumption of isotropy therefore is that the principal axes of A and B coincide.

Assume now that λ_1, λ_2 and λ_3 are *distinct*; the cases when two or more eigenvalues coincide are easier and are set later as exercises. Consider the following simultaneous equations in α, β and γ:

$$\mu_1 = \alpha + \beta\lambda_1 + \gamma\lambda_1^2$$
$$\mu_2 = \alpha + \beta\lambda_2 + \gamma\lambda_2^2 \qquad (9.27)$$
$$\mu_3 = \alpha + \beta\lambda_3 + \gamma\lambda_3^2.$$

Since μ_1, μ_2 and μ_3 are components of the tensor B, they will be functions of λ_1, λ_2 and λ_3, and hence α, β and γ are functions of λ_1, λ_2 and λ_3. By Cramer's rule, the solution of equations (9.27) for α, β, γ may be expressed as

$$\alpha = \frac{\Delta_1}{\Delta}, \quad \beta = \frac{\Delta_2}{\Delta}, \quad \gamma = \frac{\Delta_3}{\Delta},$$

where

$$\Delta = \begin{vmatrix} 1 & \lambda_1 & \lambda_1^2 \\ 1 & \lambda_2 & \lambda_2^2 \\ 1 & \lambda_3 & \lambda_3^2 \end{vmatrix} \neq 0,$$

$$\Delta_1 = \begin{vmatrix} \mu_1 & \lambda_1 & \lambda_1^2 \\ \mu_2 & \lambda_2 & \lambda_2^2 \\ \mu_3 & \lambda_3 & \lambda_3^2 \end{vmatrix},$$

$$\Delta_2 = \begin{vmatrix} 1 & \mu_1 & \lambda_1^2 \\ 1 & \mu_2 & \lambda_2^2 \\ 1 & \mu_3 & \lambda_3^2 \end{vmatrix},$$

and

$$\Delta_3 = \begin{vmatrix} 1 & \lambda_1 & \mu_1 \\ 1 & \lambda_2 & \mu_2 \\ 1 & \lambda_3 & \mu_3 \end{vmatrix}.$$

Now in §9.5, it was shown that the effect on the tensor A of the rotation of axes defined by the array (9.18) is to interchange λ_1 and λ_2; likewise the effect on B will be to interchange μ_1 and μ_2. The determinants Δ, Δ_1, Δ_2 and Δ_3 will therefore all change sign, but α, β and γ will be unchanged. By applying other suitable rotations of the axes, it follows similarly that interchanging any pair of λ_1, λ_2, λ_3 leaves α, β and γ unchanged, and hence α, β and γ are symmetric functions of λ_1, λ_2 and λ_3. From our earlier remarks in §9.3 on symmetric functions of the eigenvalues, we conclude that α, β and γ can be expressed as functions of the principal invariants of A.

In the coordinate system $Ox_1x_2x_3$, the components a_{ij} and b_{ij} of A and B are zero when $i \neq j$ and the components $a_{11} = \lambda_1$, $a_{22} = \lambda_2$, $a_{33} = \lambda_3$,

$b_{11} = \mu_1$, $b_{22} = \mu_2$ and $b_{33} = \mu_3$ are related by (9.27). Hence, relative to the axes $Ox_1x_2x_3$,

$$b_{ij} = \alpha\delta_{ij} + \beta a_{ij} + \gamma a_{ik}a_{kj} \qquad (i, j = 1, 2, 3). \qquad (9.28)$$

But this is a relation between components of second order tensors with coefficients which are invariant under a rotation of the axes, and it follows that it will hold in all coordinate systems obtained by a rotation. Hence the first part of the theorem is proved.

The converse inference, that when (9.20) holds B is an isotropic function of A, is readily proved. For, if the relation (9.20) holds in a particular coordinate system $Ox_1x_2x_3$ then on rotating the axes to new positions $Ox_1'x_2'x_3'$ it becomes

$$b_{ij}' = \alpha\delta_{ij} + \delta a_{ij}' + \gamma a_{ik}' a_{kj}',$$

since α, β and γ are invariants. Hence the functional dependence of the components of B on those of A is invariant under a rotation of the axes and B is therefore an isotropic function of A.

EXERCISES

5. In the preceding theorem, suppose that two of the eigenvalues of A are equal and the third is different. Show that, under the assumption of isotropy, two of the eigenvalues of B must then coincide, and that the components of B can be represented in terms of the components of A by a relation of the form

$$b_{ij} = \alpha\delta_{ij} + \beta a_{ij},$$

where α and β are functions of the principal invariants of A.

State also the form of the representation theorem when all three eigenvalues of A are equal.

6. In classical fluid dynamics, the components σ_{ij} of the stress tensor are assumed to be *linear* isotropic functions of the rate of strain components e_{ij}. Given that $\sigma_{ij} = \sigma_{ji}$ and $e_{ij} = e_{ji}$ for all i and j, show that

$$\sigma_{ij} = (-p + \lambda e_{kk})\delta_{ij} + 2\mu e_{ij},$$

where p, λ and μ are scalar invariants.

Remark. The student of continuum mechanics should note that this relationship can be deduced from rather more basic assumptions. If the stress components are assumed to be functions of the tensor gradient of the velocity field, then it can be shown from certain fundamental invariance requirements that there can be dependence only on the symmetrical part of this tensor gradient (which is the rate of strain tensor) and that the relationship must be isotropic. For a full discussion of this and related matters, see, for example, D. C. Leigh, Nonlinear Continuum Mechanics (McGraw-Hill, 1968), p. 145.

APPENDIX 1

Determinants

Let a_{11}, a_{12}, a_{21}, a_{22} be real or complex numbers or variables. We define

$$\begin{vmatrix} a_{11} & a_{12} \\ a_{21} & a_{22} \end{vmatrix} = a_{11}a_{22} - a_{12}a_{21}, \tag{A1.1}$$

and call the expression on the left-hand side a *second-order determinant*.
Similarly, a *third-order determinant* is defined as

$$\begin{vmatrix} a_{11} & a_{12} & a_{13} \\ a_{21} & a_{22} & a_{23} \\ a_{31} & a_{32} & a_{33} \end{vmatrix} = a_{11}\varDelta_{11} - a_{12}\varDelta_{12} + a_{13}\varDelta_{13}, \tag{A1.2}$$

where

$$\varDelta_{11} = \begin{vmatrix} a_{22} & a_{23} \\ a_{32} & a_{33} \end{vmatrix}, \tag{A1.3}$$

$$\varDelta_{12} = \begin{vmatrix} a_{21} & a_{23} \\ a_{31} & a_{33} \end{vmatrix}, \tag{A1.4}$$

$$\varDelta_{13} = \begin{vmatrix} a_{21} & a_{22} \\ a_{31} & a_{32} \end{vmatrix}. \tag{A1.5}$$

The determinants $\varDelta_{11}$, $\varDelta_{12}$, $\varDelta_{13}$ are called the *minors* of the elements a_{11}, a_{12}, a_{13}, respectively. The reader should observe that the minor of a_{11} is the second-order determinant which remains when all the elements on the same row and on the same column as a_{11} are deleted; and similarly for the minors of a_{12} and a_{13}.

The determinant, T say, on the left-hand side of (A1.2) can also be expanded, starting with the elements of the first column, giving

$$T = a_{11}\varDelta_{11} - a_{21}\varDelta_{21} + a_{31}\varDelta_{31}, \tag{A1.6}$$

where $\varDelta_{11}$, $\varDelta_{21}$, $\varDelta_{31}$ are the minors of a_{11}, a_{21}, a_{31} respectively. It is easily verified that the right-hand sides of (A1.2) and (A1.6) are identical. Expansions starting with the elements of the second or third rows or columns can also be performed, but we shall omit the details.

The reader should verify that if two rows or two columns of a determinant are equal, then the value of the determinant is zero.

Multiplication of determinants. The product of two second-order determinants is given by the following rule:

$$\begin{vmatrix} a_{11} & a_{12} \\ a_{21} & a_{22} \end{vmatrix} \times \begin{vmatrix} b_{11} & b_{12} \\ b_{21} & b_{22} \end{vmatrix} = \begin{vmatrix} a_{11}b_{11}+a_{12}b_{21} & a_{11}b_{12}+a_{12}b_{22} \\ a_{21}b_{11}+a_{22}b_{21} & a_{21}b_{12}+a_{22}b_{22} \end{vmatrix}. \tag{A1.7}$$

The rule may be verified by expanding the expressions on the two sides.

The product of two third-order determinants is given by the rule:

$$\begin{vmatrix} a_{11} & a_{12} & a_{13} \\ a_{21} & a_{22} & a_{23} \\ a_{31} & a_{32} & a_{33} \end{vmatrix} \times \begin{vmatrix} b_{11} & b_{12} & b_{13} \\ b_{21} & b_{22} & b_{23} \\ b_{31} & b_{32} & b_{33} \end{vmatrix}$$

$$= \begin{vmatrix} a_{11}b_{11}+a_{12}b_{21}+a_{13}b_{31} & a_{11}b_{12}+a_{12}b_{22}+a_{13}b_{32} & a_{11}b_{13}+a_{12}b_{23}+a_{13}b_{33} \\ a_{21}b_{11}+a_{22}b_{21}+a_{23}b_{31} & a_{21}b_{12}+a_{22}b_{22}+a_{23}b_{32} & a_{21}b_{13}+a_{22}b_{23}+a_{23}b_{33} \\ a_{31}b_{11}+a_{32}b_{21}+a_{33}b_{31} & a_{31}b_{12}+a_{32}b_{22}+a_{33}b_{32} & a_{31}b_{13}+a_{32}b_{23}+a_{33}b_{33} \end{vmatrix}. \tag{A1.8}$$

Again the rule can be verified by direct evaluation of the two sides, although the calculation is obviously tedious.

Transpose of a determinant. If the first, second, ... rows of a determinant T are replaced by the corresponding columns, the *transpose* T' is obtained. Thus, if

$$T = \begin{vmatrix} a_{11} & a_{12} & a_{13} \\ a_{21} & a_{22} & a_{23} \\ a_{31} & a_{32} & a_{33} \end{vmatrix}, \tag{A1.9}$$

then

$$T' = \begin{vmatrix} a_{11} & a_{21} & a_{31} \\ a_{12} & a_{22} & a_{32} \\ a_{13} & a_{23} & a_{33} \end{vmatrix}. \tag{A1.10}$$

It may be verified by expansion that $T = T'$.

Interchange of two rows or columns. If two rows or two columns of a determinant are interchanged, its sign is changed. Thus, if the original value is a, the value after an interchange of rows or columns will be $-a$. The reader should prove this as an exercise.

APPENDIX 2

The chain rule for Jacobians

If

$$\alpha = \alpha(u, v), \quad \beta = \beta(u, v)$$

and

$$u = u(u', v'), \quad v = v(u', v'),$$

then

$$\frac{\partial(\alpha, \beta)}{\partial(u', v')} = \frac{\partial(\alpha, \beta)}{\partial(u, v)} \times \frac{\partial(u, v)}{\partial(u', v')}.$$

Proof. Using suffixes to denote partial derivatives with respect to u, v, u', v'

$$\frac{\partial(\alpha, \beta)}{\partial(u, v)} \times \frac{\partial(u, v)}{\partial(u', v')} = \begin{vmatrix} \alpha_u & \alpha_v \\ \beta_u & \beta_v \end{vmatrix} \times \begin{vmatrix} u_{u'} & u_{v'} \\ v_{u'} & v_{v'} \end{vmatrix}$$

$$= \begin{vmatrix} \alpha_u u_{u'} + \alpha_v v_{u'} & \alpha_u u_{v'} + \alpha_v v_{v'} \\ \beta_u u_{u'} + \beta_v v_{u'} & \beta_u u_{v'} + \beta_v v_{v'} \end{vmatrix}$$

$$= \begin{vmatrix} \alpha_{u'} & \alpha_{v'} \\ \beta_{u'} & \beta_{v'} \end{vmatrix}, \quad \text{by the chain rule,}$$

$$= \frac{\partial(\alpha, \beta)}{\partial(u', v')}, \quad \text{as required.}$$

APPENDIX 3

Expressions for grad, div, curl, and ∇^2 in cylindrical and spherical polar coordinates

Cylindrical polar coordinates R, ϕ, z

$$\operatorname{grad}\Omega = \frac{\partial\Omega}{\partial R}\mathbf{e}_R + \frac{\partial\Omega}{R\,\partial\phi}\mathbf{e}_\phi + \frac{\partial\Omega}{\partial z}\mathbf{e}_z,$$

$$\operatorname{div}\mathbf{F} = \frac{1}{R}\frac{\partial}{\partial R}(RF_R) + \frac{\partial F_\phi}{R\,\partial\phi} + \frac{\partial F_z}{\partial z},$$

$$\operatorname{curl}\mathbf{F} = \frac{1}{R}\begin{vmatrix} \mathbf{e}_R & R\mathbf{e}_\phi & \mathbf{e}_z \\ \dfrac{\partial}{\partial R} & \dfrac{\partial}{\partial\phi} & \dfrac{\partial}{\partial z} \\ F_R & RF_\phi & F_z \end{vmatrix},$$

$$\nabla^2 \equiv \frac{1}{R}\frac{\partial}{\partial R}\left(R\frac{\partial}{\partial R}\right) + \frac{1}{R^2}\frac{\partial^2}{\partial\phi^2} + \frac{\partial^2}{\partial z^2}.$$

Spherical polar coordinates r, θ, ϕ

$$\operatorname{grad}\Omega = \frac{\partial\Omega}{\partial r}\mathbf{e}_r + \frac{\partial\Omega}{r\,\partial\theta}\mathbf{e}_\theta + \frac{\partial\Omega}{r\sin\theta\,\partial\phi}\mathbf{e}_\phi,$$

$$\operatorname{div}\mathbf{F} = \frac{1}{r^2}\frac{\partial}{\partial r}(r^2 F_r) + \frac{1}{r\sin\theta}\frac{\partial}{\partial\theta}(\sin\theta\, F_\theta) + \frac{1}{r\sin\theta}\frac{\partial F_\phi}{\partial\phi},$$

$$\operatorname{curl}\mathbf{F} = \frac{1}{r^2\sin\theta}\begin{vmatrix} \mathbf{e}_r & r\mathbf{e}_\theta & r\sin\theta\,\mathbf{e}_\phi \\ \dfrac{\partial}{\partial r} & \dfrac{\partial}{\partial\theta} & \dfrac{\partial}{\partial\phi} \\ F_r & rF_\theta & r\sin\theta\, F_\phi \end{vmatrix},$$

$$\nabla^2 \equiv \frac{1}{r^2}\frac{\partial}{\partial r}\left(r^2\frac{\partial}{\partial r}\right) + \frac{1}{r^2\sin\theta}\frac{\partial}{\partial\theta}\left(\sin\theta\frac{\partial}{\partial\theta}\right) + \frac{1}{r^2\sin^2\theta}\frac{\partial^2}{\partial\phi^2}.$$

ANSWERS TO EXERCISES

Chapter 1

(3) Four points, $(\pm\frac{1}{2}\sqrt{2}, \pm\frac{1}{2}\sqrt{2}, 0)$.

(4) Eight points, $(\pm 3, \pm 2\sqrt{2}, \pm 2\sqrt{2})$.

(5) Eight points, $(\pm\frac{1}{2}, \pm\frac{1}{2}, \pm\frac{1}{2})$.

(6) (i) 5; (ii) $\sqrt{61}$.

(7) $(-1, -1, 1)$; 1, 3, 0.

(8) $3\sqrt{2}$.

(10) $6/\sqrt{65}, 2/\sqrt{65}, 5/\sqrt{65}$.

(11) $135°$.

(12) $\frac{1}{3}\sqrt{3}, \frac{1}{3}\sqrt{3}, \frac{1}{3}\sqrt{3}$.

(13) 1, 1, 0.

(16) $\cos^{-1}\frac{1}{3}$.

(17) $(2, 8, -2)$.

(19) $0, \frac{4}{3}$. The points O and N coincide.

(20) $\begin{pmatrix} 1 & 0 & 0 \\ 0 & -1 & 0 \\ 0 & 0 & -1 \end{pmatrix}$; $(1, -1, -1)$.

(23) $x = x' \sin\theta \cos\phi + y' \cos\theta \cos\phi - z' \sin\phi$,

$y = x' \sin\theta \sin\phi + y' \cos\theta \sin\phi + z' \cos\phi$,

$z = x' \cos\theta - y' \sin\theta$.

(24) *Solution.* Let

$$l'_{11} = l_{22}l_{33} - l_{23}l_{32},$$
$$l'_{12} = l_{23}l_{31} - l_{21}l_{33},$$
$$l'_{13} = l_{21}l_{32} - l_{22}l_{31}.$$

Then

$$l_{11}l'_{11} + l_{12}l'_{12} + l_{13}l'_{13} = \begin{vmatrix} l_{11} & l_{12} & l_{13} \\ l_{21} & l_{22} & l_{23} \\ l_{31} & l_{32} & l_{33} \end{vmatrix}$$

$$= 1,$$

by (1.17) of the text. It follows from (1.9) that the angle between the lines with direction cosines l_{11}, l_{12}, l_{13} and l'_{11}, l'_{12}, l'_{13} respectively is zero. Hence, since l_{11}, l_{12}, l_{13} are the direction cosines of Ox' relative to the axes $Oxyz$, so also are l'_{11}, l'_{12}, l'_{13}. Thus

$$l'_{11} = l_{11}, \quad l'_{12} = l_{12}, \quad l'_{13} = l_{13},$$

as required.

The two sets of three relations similar to those for l_{11}, l_{12}, l_{13} are obtained by cyclic permutation of the first suffix in each l_{ij}. Thus

$$l_{21} = l_{32}l_{13} - l_{33}l_{12},$$
$$l_{22} = l_{33}l_{11} - l_{31}l_{13},$$
$$l_{23} = l_{31}l_{12} - l_{32}l_{11},$$

and

$$l_{31} = l_{12}l_{23} - l_{13}l_{22},$$
$$l_{32} = l_{13}l_{21} - l_{11}l_{23},$$
$$l_{33} = l_{11}l_{22} - l_{12}l_{21}.$$

(28) (a) 1, (b) -3, (c) -5.

Chapter 2

(2) (i) $(1, 2, 3)$; (ii) $(-1, -2, -3)$; (iii) $(1, 2, 2)$.

(3) $\sqrt{26}$; $\sqrt{5}$.

(4) $\frac{3}{5}$.

(5) The axes Oz', Oz coincide; Ox', Oy coincide; and Oy' is oppositely directed to Ox. $(1, -2, 2)$.

(7) $(1, 0, 0)$ and $(-1, 0, 0)$. No.

(12) (i) $(1, 0, 4)$; (ii) $(-1, 0, -4)$.

(13) The vectorial sum of the four consecutive displacements is zero.

(15) When $\hat{\mathbf{a}} = \hat{\mathbf{b}}$, or $\mathbf{a} = 0$, or $\mathbf{b} = 0$.

(17) No.

(20) $a = -1$, $b = 3$, $c = 1$.

(22) $\cos^{-1}(1/10)$.

(23) (i) $3\sqrt{2}$; (ii) 0; (iii) $-\sqrt{2}$.

(26) $\frac{1}{6}\sqrt{6}$.

(27) The resolutes are 0, 24/5 and 0.

(28) The diagonals of a rhombus intersect at right angles.

(31) (i) A vertically upward unit vector; (ii) A vector of magnitude $\sqrt{2}$ pointing downwards; (iii) A unit vector pointing NW.

(35) $\mathbf{u} = (2\lambda + 1, \lambda, -\lambda)$, where λ is an arbitrary real number.

(36) $a = b = 1$.

(37) Take $\mathbf{a} = \mathbf{b} = (1, 0, 0)$ and $\mathbf{c} = (0, 1, 0)$.

(54) $|\mathbf{F} \times \mathbf{G}|/F^2$.

Chapter 3

(1) (i) Fig. 92; (ii) Fig. 93; (iii) Fig. 94; (iv) Fig. 95; (v) Fig. 96.

(2) $C = 1$; parabola; straight line.

(3) $(1, 0, 5)$.

(7) (i) $\pi(-2\sin \pi t, \cos \pi t, 0)$, $-\pi^2(2\cos \pi t, \sin \pi t, 0)$;
   (ii) $(1, 1, e^t)$, $(0, 0, e^t)$;
   (iii) $t > 0$: $(1, 1, 0)$ and $(0, 0, 0)$;
     $t < 0$: $(-1, 1, 0)$ and $(0, 0, 0)$.

(8) $(e^{-t}\cos t, e^{-t}\sin t, 0)$. See Fig. 97.

(13) $\mathbf{a}$ and $\mathbf{b}$ are such that $\mathbf{a} \times \mathbf{b}$ is a constant vector; and so $\mathbf{a}$ and $\mathbf{b}$ are parallel

to a fixed plane and are such that the parallelogram with edges **a** and **b** has constant area.

(14) (i) See Fig. 98. This is a smooth, simple closed curve. (ii) See Fig. 99. This is a smooth curve.

(16) See Fig. 100.

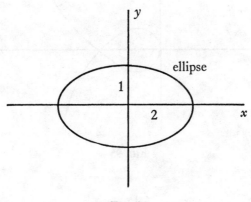

FIG. 92

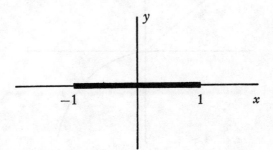

FIG. 93

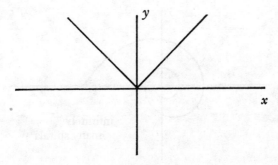

FIG. 94

(17) See Fig. 101.

(18) $\mathbf{r} = (a \cos(s/c), a \sin(s/c), bs/c)$, where $c = (a^2 + b^2)^{\frac{1}{2}}$.

(23) $\frac{1}{2}$.

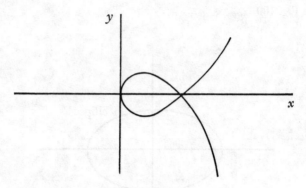

Fig. 95

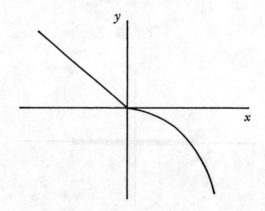

Fig. 96

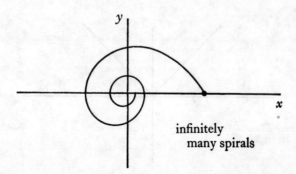

infinitely
many spirals

Fig. 97

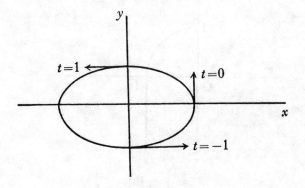

FIG. 98

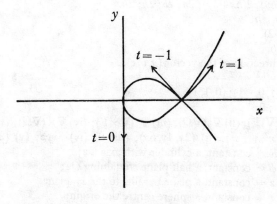

FIG. 99

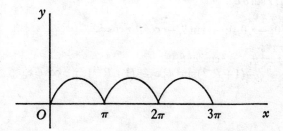

FIG. 100

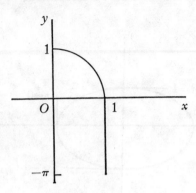

FIG. 101

Chapter 4

(1) Open.

(3) $2(ax+hy)$; $2(hx+by)$; $2a$; $2b$; $2h$; $2h$.

(5) $x/r, y/r, z/r$.

(9) $(-2, 0, 2)$.

(14) $27/\sqrt{6}$.

(15) The direction of the vector (a, b, c).

(16) $x/a^2, y/b^2, z/c^2$.

(18) 2; $(-1, 0, -1)$; $(0, 1, 1)$.

(19) $2+6z$; 0.

(23) (i) $\nabla(\nabla \cdot \mathbf{F})$; (ii) $\nabla \cdot (\nabla\Omega)$; (iii) $\nabla \cdot (\nabla \times \mathbf{F})$; (iv) $\nabla \times (\nabla\Omega)$; (v) $\nabla \times (\nabla \times \mathbf{F})$.

(25) (i) $2(x^2, y^2, z^2)$; (ii) $(2x, 4y, 6z)$; (iii) $\mathbf{0}$; (iv) $3xyz$; (v) $(2, 2, 2)$; (vi) $\mathbf{0}$.

(33)   (i) $R = $ constant, a cylinder with axis Oz;

        $\phi = $ constant, a half plane containing Oz;

        $z = $ constant, a plane parallel to the xy-plane.

   (ii) $r = $ constant, a sphere centre the origin;

        $\theta = $ constant, an infinitely long cone with vertex O, axis Oz;

        $\phi = $ constant, a half-plane containing Oz.

(36) $h_1 = h_2 = (\sinh^2 \xi + \sin^2 \eta)^{\frac{1}{2}}, h_3 = 1$; ellipses, hyperbolas.

(39) $R\mathbf{e}_\phi$; $r\sin\theta\,\mathbf{e}_\phi$.

(40) $R\mathbf{e}_R + z\mathbf{e}_z$.

(41) $1, 1, 0$; $-r, 0, 0$; $-r\sin\theta, -r\cos\theta, r\cos\phi$.

(43) $\mathbf{e}_R + \frac{1}{2}\mathbf{e}_z$.

(46) $-r^{-2}\cot\theta, -r^{-2}\operatorname{cosec}^2\theta, 0$.

(51) $\{R\cos^2\phi - (1+R^{-1})\sin^2\phi\}\mathbf{e}_R + (1+R^{-1})\sin\phi\cos\phi\,\mathbf{e}_\phi$.

Chapter 5

(1) $\frac{4}{5}\sqrt{5}(2+\pi^2)\,a^2$.

(3) $\frac{1}{2}63\sqrt{3}$.

(4) πa^3.

(5) $\frac{1}{12}(5\sqrt{5}-1)+\frac{1}{3}\sqrt{3}$.

(6) (i) $7/6$; (ii) $\mathbf{0}$; (iii) $\left(\dfrac{3}{4}, \dfrac{8\sqrt{2}}{15}, \dfrac{5}{12}\right)$.

(7) $2\pi c(a, 0, \pi c)$.

(8) $-2\pi a$.

(9) 6π.

(10) $2\pi a(0, b, a)$; $2\pi^2 b(a^2 + b^2)^{\frac{1}{2}}\mathbf{k}$.

(14) $\frac{1}{3}$.

(15) $\frac{1}{2}(3e - 7)$.

(17) $4(3 - \sqrt{2})/21$.

(18) $\frac{1}{2}\pi(b^4 - a^4)$.

(19) 8π.

(20) $-\frac{1}{2}\pi$.

(21) $\frac{1}{4}\pi a^3 b$.

(23) $\frac{1}{16}$.

(24) $\pi^{\frac{1}{2}}$ (see Example 19, § 5.7).

(25) $\frac{1}{6}$.

(26) $\frac{4}{15}\pi abc(a^2 + b^2 + c^2)$.

(28) $\frac{1}{2}(\sqrt{3}\cos\phi, \sqrt{3}\sin\phi, -1)$; $2r\,dr\,d\phi$.

(29) $2\displaystyle\int_{0}^{2\pi} (a^2\sin^2 u + b^2\cos^2 u)^{\frac{1}{2}}\,du$.

(30) $\frac{1}{6}\pi(5\sqrt{5} - 1)$.

(31) $z = \{3(x^2 + y^2)\}^{\frac{1}{3}}$.

(33) $\frac{1}{8}$.

(34) $6\pi a^3$.

(35) (i) $\frac{1}{2}\pi$; (ii) $(0, 0, \frac{4}{3}\pi)$.

(36) (i) $(\frac{1}{2}, \frac{1}{2}, 1)$; (ii) 1; (iii) $(\frac{1}{2}, -\frac{1}{2}, 0)$.

(37) $(0, 0, -\frac{1}{9}\pi)$.

(38) 54.

(39) $4\pi\{(a^2 + b^2)^{\frac{5}{2}} - a^5 - b^5\}/15b$.

(41) $\pi a^5/15$.

(42) $\frac{2}{3}$.

Chapter 6

(1) $12\pi a^5/5$.

(3) $\mathbf{F}$ continuously differentiable in τ; and τ bounded by a simple closed surface S.

(15) $\mathbf{F}$ continuously differentiable on S; and S is a simple open surface spanning the correspondingly oriented curve $\mathscr{C}$.

Chapter 7

(1) (i) 2; (ii) 1; (iii) 4.

(4) $-\dfrac{1}{r}$; $\frac{1}{2}r^2$.

(6) $\mathbf{A} = (0, \frac{1}{2}x^2 - yz, -xz)$.

(7) $\psi = -\cos\theta$.

(14) $\frac{2}{3}\pi$.

Chapter 8

(1)
$$\begin{pmatrix} a_{22} & -a_{21} & a_{23} \\ -a_{12} & a_{11} & -a_{13} \\ a_{32} & -a_{31} & a_{33} \end{pmatrix}$$

(12)
$$\begin{pmatrix} 0 & b_3 & -b_2 \\ -b_3 & 0 & b_1 \\ b_2 & -b_1 & 0 \end{pmatrix}$$

Chapter 9

(1) $c_1(1, 0, 0)$, $c_2(0, 1, -1)$, $c_3(0, 1, 1)$, where either $c_1 = 1$ and $c_2 = c_3 = \pm\frac{1}{2}\sqrt{2}$, or $c_1 = -1$ and $c_2 = -c_3 = \pm\frac{1}{2}\sqrt{2}$.

(2) (i) $I_1 I_3$; (ii) $I_3(I_1^2 - 2I_2)$.

(3) $\mathcal{J}_1 = \lambda_1 + \lambda_2 + \lambda_3$, $\mathcal{J}_2 = \lambda_1^2 + \lambda_2^2 + \lambda_3^2$, $\mathcal{J}_3 = \lambda_1^3 + \lambda_2^3 + \lambda_3^3$.

(4) $n = 0$ and $n = 2$.

(5) $b_{ij} = \alpha\delta_{ij}$.

GENERAL INDEX